Body and Awareness

Ways of Being a Body ~ Volume 3

Body and Awareness

edited by

Sandra Reeve

Published in this first edition in 2021 by:

Triarchy Press
Axminster, UK

www.triarchypress.net

A catalogue record for this book is available from the British Library.

ISBNs:
Print: 978-1-913743-00-0
ePub: 978-1-913743-01-7
pdf: 978-1-913743-02-4

Front cover image: Greta Berlin ~ www.gretaberlinsculpture.com

tp

Dedicated to:
Suprapto Suryodarmo
(1945-2019)

"A world without words is still in communication,
but there are no words."
(Suprapto Suryodarmo, Bali, 2014)

Contents

Introduction

I am always interested in the spaces between.

In this book I have brought together an interdisciplinary collection of viewpoints and practices on the topic of Body and Awareness. Each author was asked to offer their own definition of awareness and to write from their own somatic/movement practice. This has created an anthology of stimuli, perceptions and experiences and a wide knowledge base to savour, to question, to explore further and a wonderful resource for those who seek to find "a pattern of patterns of connections" (Bateson, 1979: 18)[1].

Back in 1979 Gregory Bateson maintained, for example, that somatic changes, provoked by the environment or by physiology, could precede a genetic shift, but that the genetic disposition at any given moment would also limit the adaptive potential of the organism (ibid: 178). This analysis brings together choice and a sense of proportional self-regulation of the system. These principles are clearly visible in movement. We can consciously adapt and transform how we move. On the other hand, our movement capacity is also limited by the characteristics of our own particular bodies, partially inherited, partially developed through our own environmental conditionings and presently influenced by our health.

I invite you to find your own pathway through these widely varied chapters, which can be read in any order. Follow your impulse. You can move between 'body and awareness' within a therapeutic environment or as perceived within shamanic, healing and mindfulness traditions. You can savour 'body and awareness' within a variety of performance contexts, within different somatic, walking and movement practices, within an arts-based initiative for children and within an autobiographical creative arts enquiry. You can experience 'body and awareness' through movement and poetry as well as pondering its nature through an embodied philosophical exploration of body, awareness and the imagination.

Somewhere between them all you may find the words to share your own experiences of the non-verbal realms to enrich each of our ways of facing the unknown.

Sandra Reeve ~ *Westhay, November 2020*

---

[1] Bateson, Gregory (1979) *Mind and Nature*, Hampton Press

The Actively Imagining Body

Body, Awareness and Active Imagination – Transformative Learning in Primary Education

Helen Edwards

Abstract

This chapter explores ways in which awareness is seen through the lens of the actively imagining body. Some thought is given to the development of sensory motor competence and the kinaesthetic basis of imagination. Vignettes are chosen from Story Makers Project, an arts-based initiative for children in Primary Education needing additional support with communication, and the adults working with them. The vignettes illustrate the immersion of the body in artistic process, the development of awareness and how this leads to engagement with imaginative, unconscious life.

Approach to Awareness

For the actively imagining body, awareness is approached through the senses in structured ways. Sensory motor skills enable a developing child to learn about and control their body. As the child learns to move through their environment, gathering sensory information and practising skills, the motor system learns to receive, interpret and respond successfully to that information. Sensations of movement and gravity, from muscles and joints, are integrated into developing postural security, motor planning, awareness of both body sides and orientation in space. These new experiences enable perceptual motor development, that is to say, more complex body schema: eye-hand coordination, visual-spatial perceptions and auditory-language skills. Developmental theorists propose that sensory motor integration – communication and coordination between the sensory and motor system – is the foundation for cognitive development, intellect and daily living skills. Bruner (1965) described a continuous development, whilst Piaget (1936) defined age-acquired competencies.

System	Location	Provides information about:
Tactile	Skin	The environment and object qualities (touch, pressure, hard, soft)
Visual	Retina of the eye	Objects and persons, spatial recognition and movement in the environment (high, low, near, far)
Auditory	Inner ear	Sounds in environment
Gustatory	Chemical receptor in tongue	Different types of taste (salty, sweet, sour)
Olfactory	Chemical receptor in nasal structure	Different types of smell (flowery, acrid, putrid), Important in bonding with mother and others
Vestibular	Inner ear	a) Where the body is in space and whether or not we or our surroundings are moving b) Maintaining balance and equilibrium, interpreting the force of gravity and speed/quality of movement
Proprioceptive	Muscles, joints, tendons; receptors monitoring contraction/stretching of muscles, bending/straightening, pulling/compression of joints	a) Where a body part is and how it is moving b) Motor control and movement c) Maintenance of muscle tone and tension and smooth motor control

Table 1: *Location and functions of the sensory systems* (Tarakci & Tarakci, 2016)

The actively imagining body approach places attention on sensory motor processing and integration. Problems processing sensory information that the brain receives, especially proprioceptive and vestibular, may affect motor skill development and underlie learning and communication difficulties. Children with poor proprioception (inability to locate and control body parts) and vestibular issues (inability to orientate and balance the body in space and gravity) may feel off-balance and out of control. They may withdraw from playing with objects and exploring movement which can then impede motor skill development.

Story Makers takes inspiration from sensory engagement with world-renowned collections and unique gallery spaces in Oxford Museums. In 2018 pupils from three Oxford Primary School groups visited musical instrument collections in the Ashmolean Museum of Art and Archaeology. Each school group (ten children, two adults and three volunteers) learnt about developments in musical instrument-making brought about by significant technological advancements in the Renaissance. The collections include the Stradivarius Messiah Violin and Kirkman Harpsichord.

Vignette 1

In the Education Studio a violin-maker demonstrates his tools and shares experiences of making violins and restoring the Messiah violin. Children handle the beautifully crafted violin parts and excitedly point to 'back' and 'neck' as we learn of their correspondence with the human body. We examine the wood and think about the Tyrolean spruce used in Renaissance violin-making. We reconstruct a violin from its parts and grasp how the instrument is made whole.

In the Music and Tapestry Gallery participants use their senses to investigate the gallery, the cabinets of violins and wall-hung tapestries, finally coming to rest on the floor. I invite the violin-maker to play his cello. As we listen to the musical sounds we notice ways in which their resonance and movement engage the body, sensation, feeling and any images that arise in the imagination.

We begin with games guessing what animal sounds the violin-maker is playing on the cello and go on to suggest animals for him to play and decide if they seem to correspond. Next are games naming feelings/emotions played on the cello and subsequently suggestions for emotions to be played, with participants discussing whether the sound does indeed resonate with the emotion. We build short sequences linking some of these sounds together as preparation for listening to longer passages of Bach's Cello Suite No 1 in G; Prelude, Minuet I and II and Gigue.

The focus on self-awareness continues as we listen to an introduction to Prelude with attention to feeling and sensation. Participants lie down and I suggest body parts to notice (head, neck, arms, fingers, legs, etc.) to support growing awareness of the body; as well as shape, position, contact with floor and sensations of gravity. I ask everyone to make themselves comfortable and notice their breathing. As the music begins, I invite participants to visualise a beautiful landscape and notice changes in sensation, feeling and images arising. I encourage participants to be patient and remember everyone's imagination will be different. As Prelude continues, I invite participants to enter and explore their landscape, noticing plants, animals and birds, and any possessions or people they would like with them. As Prelude ends, I invite everyone to say goodbye to their landscape and become aware of their body and the gallery space.

Participants now sit in pairs and introduce Story Makers' dolls they had made earlier to each other. During the Minuets and Gigue they invite the dolls to dance whilst remembering their visualised landscapes. During Minuet I, one partner's doll is leader and the other follows. For Minuet II they change over. During Gigue the dance emerges and is co-created by the dolls as they mix and blend movements. Afterwards everyone has a large piece of paper and coloured pastels and creates a picture of their landscape adventures.

Sensation, Movement, Body Image and Intentionality

Vignette 1 illustrates the immediacy of engaging the senses and the imagination. In the Education Studio, participants use gesture and sight to identify correspondences between the parts of a violin and the human body, for example the neck, waist, finger-board and tail-piece by making the shapes of each of the parts of the violin in the corresponding part of the human body and showing each other. Faint wooden sounds arise from handling the violins, aromas of wood-varnish and resin stimulate hearing and smell. Walking to and around the gallery engages body movement with sensory experiencing. The initial tight energy of nervously clumping participants relaxes into a flow of individuals moving with awareness of their presence in the gallery, orientating themselves and noticing each other. The group recognise their collective whole as they form a circle sitting by the Messiah violin.

Games where the violin-maker plays the cello encourage people to share sensations and feelings arising in response to his sounds, using gesture and word to communicate personal metaphors and images. All discover and learn together, naturally taking turns and listening to one another.

Fig. 1: *Images of Sound*

"Music sounds like people walking, planting seeds in the earth. A rhythm is beating of footsteps and horses' hooves. Everyone stands in a circle, holding hands and dancing happily, people and animals together."   (Story Maker, age 10)

"Prelude is like being scared, nervous and unsure of what will happen. It's like sensing something change – a season – late summer. It's like time changing and life going in a new direction. Prelude is like going out in nature, warm, sunny, rain then sun again, weather changing like a story starting to end. I feel hot, the wind blows on my face and feelings and emotions are starting and colliding."   (Story Makers' groups)

Immersion in music, movement and gesture, encourages individuals to pay attention to themselves, be conscious of the continuous nature of their self-experience, inner attention, and feel their intimate encounter with self.

Fig. 2: *Story Makers' Drawing: 'The Experience of Music'*

The visualisation of a personal landscape becomes the medium through which a fantasy image may arise from the imagination in an active encounter with unconscious experience. This is illustrated in the words of one Story Maker, age 8:

> At first I feel awkward, but realise I am OK when I see everyone is lying down. I feel relaxed and happy. In my imagination I find a wonderful place to play with a friend, it is the summer in my world, the summer holidays.

The actively imagining body approach to awareness illustrated here is phenomenological, providing direct description of human experience and perceptual contact with the world. Whole, here-and-now self-awareness may offer a non-judgemental lens "engaging body, mind and spirit, psyche and soma, that offers a body and awareness approach and may bring organismic self-regulation" (Yontef, 1993). A person may understand their beliefs, needs, strengths and creativity and find and regulate new ways of living. Story Makers' activities value body, sensation and feeling as integral to imaginative activity and learning about self and others. All aspects of being human are considered related, the mind as part of the body, the body affecting the mind.

Merleau-Ponty's philosophy posits body image as an awareness of being situated in the world, with roots to a person's sense of "interpersonal corporeality with other". He suggests that the body's kinaesthetic sense of its movement and existence becomes consciously known through a pre-conscious system of bodily movements and spatial equivalences situating a person within human relatedness to self, other and nature. He suggests bodily movements have "an 'intentional arc' which projects around about us our past, our future, our human setting ... which results in our being situated in all these respects" (Merleau-Ponty, 2012). Reeve describes intentionality as "a way in which consciousness is always pointing towards something" (Reeve, 2011). I will expand on this prospective element of intentionality later with reference to active imagination.

The actively imagining body approach also draws on capacity for reflection on subjective experience (thoughts, feelings, sensation/process) and how reflexive capacity may reveal hitherto unconscious intentionality, thus offering existential awareness. Body image is fleshed out through developing awareness, the kinaesthetic sense and symbolic artistic imagery.

Thus bodily space reflects an original form of intentionality – a precognitive encounter with the world as meaningfully structured, and the embodied image is closer to the symbolic, bringing conscious and unconscious elements together; "the embodied picture transcends what is consciously known" (Schaverien, 1992).

Moving with awareness of the body may reveal a new experiential understanding of a person's sense of self. In fostering awareness in movement of fixed thoughts and emotions, which may inhibit the ability to be present and engaged, embodied movement may create room for deepened engagement with new possibility. The embodied image offers transcendence to a more dynamic state of awareness and capacity to be more fully present, engaged and responsive to what is presenting itself.

Vignette 2

> *In the European Gallery a harpsichord-maker tells Story Makers about harpsichord-making. He plays two dances on the Kirkman Harpsichord. Initially participants lie on the floor listening to Handel's Allemande in F major HWV 476. With closed eyes, all become aware of body, breath and floor, travelling into the music, letting sensation, feeling and images arise. Next, participants dress up in fine-coloured cloths and choose a peacock feather and dance partner. Under the high-ceilinged gallery and vast Baroque portraits steeped in history, participants stand in pairs noticing their feet on the ground, body posture and their position in the group and gallery space. I name and describe the vertical and horizontal axes the floor and walls give us and invite everyone to hold out their arms horizontally with feathers vertical and notice any sensations of gravity in their bodies. As Allemande is played again participants move freely in pairs, however they like, letting their feathers dance. I invite everyone to discover their own unique way of moving and imagining being light as a feather. We begin and end dances with a bow or curtsey. One partner begins leading, with their partner following, then we swap.*
>
> *The second piece, Arne's Presto, Sonata no 6 in F minor, is a jig of different tempo, key and feeling. This time the dance arises spontaneously between partners. Afterwards I invite participants to talk together about the thoughts, feelings and emotions they had noticed arising during the music. We imagine stories about characters in the gallery's portraits around us, stimulated by physically enacting their body postures and gestures. Next, participants make large drawings of their experiences using coloured pastels on paper.*

Engagement as Artists in the Story Makers' Group Process

Body, breath, voice, sound, resonance, gesture, movement and awareness incubate together in the unique space of each gallery. A nest of sensory experience is woven in which one child curls up momentarily on feeling a need for comfort and safe withdrawal while an adult rests in stillness. Another child feels alone and stamps her foot when the music is introduced. Her foot-stamping marks musical tempo and communicates her participation. An adult reacts, about to intervene and stop her, but as

I beat musical time the adult becomes aware of the girl's attunement to, and involvement in, the music. The adult's recognition of the girl's actions as participatory, rather than disruptive, stop her from interrupting the child. On realising herself as participating the girl adapts, quietening her stamping. She joins in the visualisation, embedding new participation and cohesion within her self-experience. The structured exercises shaping the sensory nest enable the fledgling participants to enter the research together as artists, experiencing music and its resonance in the body-psyche. This approach enables Story Makers to transcend traditional educational roles. Children and adults actively engage together as artists.

In the 1930s, Russian psychologist Lev Vygotsky posited that "experience of cooperative efforts in the group may serve as a representation of inner cooperation and wholeness which may become internalised within the individual self." (Vygotsky, 1997). His ideas included a zone of proximal development – "distance between the actual developmental level as determined by independent problem-solving and the level of potential development as determined through problem-solving under adult guidance, or in collaboration with more capable peers" (ibid). Wood developed Vygotsky's social learning theories, introducing the term 'scaffolding', a process offering a "structure of interaction between an adult and a child with the aim of helping the child concentrate on the difficult skill she is in the process of acquiring" (Wood, 1976).

Our approach 'scaffolds' learning for adults and children. The structured yet experimental processes promote development of body-mind connection, engagement of the nervous system and sensory integration. Participants relax at the reassuring sight and sound of others settling on the gallery floor. They become aware of one another's continued efforts to create something together, and recognise new aspects of self-encounter, inner cohesion and identity. They may become conscious of the group cooperation and shared endeavour and begin to internalise such cohesion as inner wholeness. Thus the actively imagining body approach scaffolds sensory motor development with its vital role as an integral foundation for active imagination.

Visualisation, free movement and drawing offer active encounter with the unconscious in a process Jung (1970) called 'active imagination'. This involves letting a fantasy image from the unconscious come into the field of inner attention, giving the fantasy some form for free association to discover its nuances and unconscious roots, and next confronting any ethical or moral dilemmas arising from the new information. Jung described active imagination as possible in a number of ways, e.g. dance, art, drama, and as a "natural, inborn process, a movement out of the suspension between two opposites, a living birth that leads to a new level of being, a new situation" (ibid).

> I am learning that if I listen, my imagination has so much to say.
> My art never stops I can always be creative. (Story Maker, age 8)

The active imagination process supports new levels of being to arise for participants, a new dialectic position in the active encounter between conscious and unconscious experience, placing agency within the individual. Jung called this psychic process 'the transcendent function' (Jung, 1956). He believed that there is a 'prospective' element to the fantasy image, that "purposively interpreted it seems like a symbol" (Jung, 1971), a kind of intentionality, as I mentioned earlier in relation to bodily movement seeking to transcend identity and reveal the scope of the individual from the collective matrix.

Relating to objects in museums brings an unconscious experience of these collective symbols of learning and transformation into dialogue with conscious thought. In free association (or "phantasy thinking"), "directed thinking is brought into contact with the oldest layers of the human mind, long buried beneath the threshold of consciousness. The phantasy products engaging the conscious mind are, first of all, waking dreams or daydreams." (Jung, 1956).

An invitation to enter Baroque music with inner attention brings discovery of fantasy image. New motifs and associations arise through visualisation, free dance and drawing. The sequences offer scaffolding for emerging reliable experiences of self through a range of sensory pathways and support sensory motor integration. Presence-openness allows felt meaning to emerge out of experiential immersion in expressive patterns of the symbolic medium. The imagination may combine feelings, sensations, memories and intuitions in symbolic images which are original, deeply personal and meaningful.

The Role of Awareness and Expressive Movement in Active Imagination Processes

Interplay of music and movement stimulates active multi-sensory learning. Sound awakens the muscular and nervous system as well as developing the ear. Rhythm, harmony and melody are felt in movement sensations, enabling understanding of nuances of music through the body.

Abramson formulates it as follows: "Whenever the body moves, the sensation of movement is converted into feelings that are sent through the nervous system to the brain which, in turn, converts that sensory information into knowledge. The brain converts feelings into sensory information about direction, weight, force, accent quality, speed, duration, points of arrival and departure, straight and curved flow paths, placement of limbs, angles of joints, and changes in the center of gravity." (Urista, 2016). Thus moving with music may enable access to the kinaesthetic sense in the human experience of understanding qualities and rhythms in musical sounds.

The structured movement exercises support awareness, self-expression and intimate relationship with self. Using Story Makers' dolls to improvise and choreograph movement offers rehearsal and mapping of sensory pathways. Tensions, restrictions, arcs, lunges and directions rehearsed in doll dances take stage in the body in the feather dances. The sequence of leading, following and then improvising together enables growing awareness of body movement and how it is informed by inner impulse, music, feather, costume, partner or gallery-portraits. The drawing process that follows with its emphasis on mark-making on paper's surface may portray the learning coming to awareness through movement, as qualities felt in the movement are made visible to the conscious mind through the substantiality of the drawn marks.

In the dances, participants are encouraged to move and improvise from feeling. Movement focus is on the creative experience of music, rather than a choreography. In Vignette 2 everyone is asked to notice the orientation of their body in space, gravity and direction, the nature and the extent of their movements. This invites participants to awaken their subjective awareness of their own body. Whitehouse, a movement pioneer with Jungian training, believed that "the kinaesthetic sense is not always conscious but could be awakened through the subjective connection" (Whitehouse, 1999). Story Makers are invited to follow sensations allowing inner impulse to lead. This engages the proprioceptive and vestibular senses and promotes psychomotor integration.

An adult commented, "I didn't believe I could dance. Once I started moving so much came out of me. Then the dancing really started. I drew the picture [Figure 3] of myself in the harpsichord making the music dance. I am a dancer."

The awareness of 'what I am doing' coming together with 'what is happening to me' brings these opposites together and identity is transcended. Whitehouse describes this process:

'I move,' is the clear knowledge that I, personally, am moving. The opposite of this is the sudden and astonishing moment when 'I am moved'. It is a moment when the ego gives up control, stops choosing, stops exerting demands, allowing the Self to take over moving the physical body as it will. It is a moment of unpremeditated surrender that cannot be explained, replicated exactly, sought for or tried out.   (Whitehouse, 1999).

Fig. 3: *Dancer in Harpsichord*

A new vision of self can be glimpsed.

The conscious experience of physical movement producing changes in the psyche is an important idea, central to the actively imagining body approach.

A person's posture and movement may reflect his 'stance' towards life and others, based on body memory of repeated emotional or sensory experience. In paying attention to bodily sensations, posture and movement, unconscious material, forgotten memories and associations can be accessed. A change in physical movement can prepare a person for a change in his emotional state.   (Hendry, 2018)

Body movements, gestures and images arising for Story Makers hold conscious and unconscious attitudes, a condensed expression of the whole psyche. The non-verbal, dancing dialogues explore polarities of 'mover' and 'being moved'. A new dialectic position is established enabling opposite attitudes to engage in mutual influence in the final co-created dance. The body and awareness work is a foundation for learning.

Two individual cameos arise from the vignettes which illustrate the transformative learning and offer examples of the role of language and words in the unfolding stories.

Individual Cameo

One girl, described as selectively mute, discovers she enjoys following and leading with her partner in the doll dances. Initially slow to engage, she becomes delighted to discover freer, less controlled, movements shared confidently in final co-created movements. In the feather dances, the girl clutches her feather and makes small off-balance, stilted movements in following her partner. She smiles when they curtsey. She begins leading but stumbles. Realising her partner is waiting she rebalances and continues leading. She waves her feather with confidence and joy. The group see and welcome her enjoyment, later she is more included in activities. She enjoys the drawing and writes a story, Mountain Journey, which she, and the group are transfixed by. The story illustrates her creativity and transformation in confidence.

Mountain Journey

When it goes dark, a girl is alone and scared, fending off dark animals. She finds a door in the mountain which leads to the top. There she finds a golden recorder.
When she plays the recorder music like melted gold comes out, the light returns and light animals come running to her. She realises she is made of gold and she is ok. The girl realises she is a musician.

Individual Cameo

As the feather dances start, a child begins to withdraw, sitting down and looking away. After a time he glances round; seeing everyone dancing, he crawls closer. He stops by me, a short distance from the others. I crouch down. He takes the feather from my hand and uses it like a fencing épée. These spirited movements are staccato, chaotic. He stands up and moves in small staccato steps. He passes me the feather gently. I hold and waft it, first staccato then softly. I pass it back gently. As he senses my sensing his interest in the group, through gesture, we move towards the others. He draws a still figure watching other dancing figures, a harpsichord and a large peacock feather.

Figure 4: *Figures, Harpsichord and Peacock Feather*

He writes a story about a boy going to a musical instrument competition to play the harp on a big stage. He doesn't think he will get there. He finds a rowing boat. He discovers the boat is tied to the competition stage so he can pull himself there. This story helps him find language to articulate his feelings, the old belief he can't have good experiences and is alone and the learning that he can be supported in relationship.

The Moment of Language

> Non-verbal experiences registered in sensory, kinaesthetic and
> somatic modalities become organised into non-verbal
> perceptual images, modes which occur before symbolisation,
> enabling categorisation of experiences, object and events into
> prototypic images which may then be the basis for connection
> to linguistic expression. (Chaiklin, 2009)

The body and awareness processes key to this active imagination approach
are central to engaging imaginative and unconscious life. Visualisation,
expressive movement and drawing enable unconscious content to move
towards consciousness. The fantasy images arising offer the "highest
expression of a person's individuality and may even create that
individuality by giving perfect expression to its unity." (Jung, 1971).
Stories emerge from the images and convey aspects of the dialectical
encounter in which old positions in consciousness and unconsciousness
are transcended. The stories enrich language, bring new understanding
and offer an important forum for discovery and integration of "...how the
ego will relate to this position, and how the ego and the unconscious are
to come to terms" (Jung, 1971). Individuals' identities, relationships and
worlds are expanded as language is excavated and discovered.

Conclusion

The actively imagining body approach offers a medium that encourages
awareness and directs the engagement of the body and mind towards
transformative learning. Through this lens new aspects of self may come
into view and, as these become conscious, change and adaptation to
identity and experience of relatedness to other and the world become
possible.

The Story Makers' vignettes illustrate how children and adults in
group situations value the range of senses and emotions needed to help
them build self-esteem, self-confidence, emotional expression and
communication skills. They may develop and sustain closer relationships
with each other and their peers, trust in each other and hope for the
future. Adults develop awareness of changes they experience in
themselves and changes they witness in the children. The transformation
in the thoughts, feelings and beliefs the adults hold about the young
people they work with will have enduring impact on their practice as

educators. The effect of directed-rational approaches, demarcations of authority and fixed teacher/student roles may become apparent as barriers to engagement in relational learning. Movement and play are experienced as well-researched effective tools for learning and self-development in education.

Innovative images and stories arising through imaginative processes portray moving experiences of transformation within the personality arising from engagement in unconscious life. Stories scribed from images and experiences poetically describe learning. Children and adults find words capturing journeys of discovery and change. The transformative experiences enable participants to enter new aspects of identity and relatedness and these deepened levels of engagement emerging through the Story Makers' process may sow seeds for the blossoming of new imaginal dimensions to relational learning for adults and children in primary education.

Actively imagining body approaches can support the role of imagination in our lives which:

> ...may be awakened and nourished through attention to the present, feeling world of the body and to whatever appears as we make. In this way we enter into the poetics of our experience.
>
> (Tufnell and Crickmay, 2004)

Acknowledgements

Thanks to Fusion Arts, a Community Art Agency in Oxford, with whom I have worked in partnership to conceive and develop Story Makers since 2010. Also thanks to The Ashmolean Museum of Art and Archaeology, Wood Farm, Bayards Hill Primary Schools, John Henry Newman Academy School and to funders Children in Need.

References

Bruner, J. (1965) 'The Growth of Mind', *American Psychologist*, 20(12), 1007-1017

Chaiklin, S. and Wengrower, H. (2009) *The Art and Science of Dance Movement Therapy.* Routledge

Chodorow, J. (1991) *Dance Therapy and Depth Psychology: The Moving Imagination.* Routledge

Hendry, A. and Hasler, J. (2018) *Creative Therapies for Complex Trauma.* Jessica Kingsley

Jones, R. Clarkson, A. Congram, S. Stratton, N. (2008) 'The Dialectical Mind in Education and the Imagination', *Education and Imagination, Post Jungian Perspectives.* Routledge

Jung, C. G. (1916/1970) *The Structure and Dynamics of the Psyche. The Collected Works of C.G. Jung.* Routledge

_____ (1956) *Symbols of Transformation. The Collected Works of C.G. Jung.* Routledge and Kegan Paul

_____ (1971) *Psychological Types. The Collected Works of C.G. Jung.* Routledge and Kegan Paul

Nielson, T. *et al.* (2010) *Imagination in Educational Theory and Practice: A Many-Sided Vision.* Cambridge Scholars Publishing

Merleau-Ponty, M. (1945/2012) *Phenomenology of Perception.* Routledge

Piaget, J. (1936) *La Naissance de l'Intelligence chez l'Enfant.* Delachaux et Niestlé

Schaverien, J. (1992) *The Revealing Image: Analytical Art Psychotherapy in Theory and Practice.* Jessica Kingsley

Tarakci, D. and Tarakci, E. (2016) 'Growth, Development and Proprioception in Children' in D. Kaya (ed.) *Proprioception – The Forgotten Sixth Sense.* OMICS Group International

Tufnell, M. and Crickmay, C. (2004) *A Widening Field: Journeys in Body and Imagination.* Dance Books

Urista, D. J. (2016) *The Moving Body in the Aural Skills Classroom: A Eurythmics Based Approach.* Oxford University Press

Vygotsky, L. S. (1997) *Vygotsky – Collected Works of L.S. Vygotsky*, ed. R.W. Rieber. Plenum

Whitehouse, M. S. (1958/1999) 'The Tao of the Body' in P. Pallaro (ed.) *Authentic Movement: A Collection of Essays by Mary Starks Whitehouse, Janet Adler and Joan Chodorow* (pp. 41-50). Jessica Kingsley

Wood, D., Bruner, J. and Ross, G. (1976) 'The Role of Tutoring in Problem Solving', *Journal of Child Psychology and Child Psychiatry*, 17, 89-100

Yontef, G. (1993) *Awareness, Dialogue and Process.* The Gestalt Journal Press

---

Helen Edwards has always loved nature: gardening, swimming, dancing, walking and painting outdoors since a child. She was very naturally drawn to study for a Masters degree in Integrative Arts Psychotherapy in the late 1990s and since to train in ecotherapy and in movement-based approaches which place the body in an ecological setting – Butoh and Amerta Movement. She has developed innovative ways of sharing creative work in schools, hospitals, community and outdoor settings alongside private practice work in Oxford.

hels19@hotmail.com

The Alchemical Body

Movement and the Way of the Shaman

Hilary Kneale

Abstract

This chapter introduces the idea that, while embodying a 'shamanic state of consciousness' within the nature of what is called the Alchemical Magician, a shift of perception from ordinary to non-ordinary reality can arise. In shamanic terms, alchemy (or the Alchemical Magician) "symbolically ... transforms lead into gold; heavy mental and emotional states can become ones that are light and bright" (Luttichaü, 2017 : 145).

The chapter's title acknowledges the coming together of two streams of practice, shamanism and environmental movement, which combined create a form of alchemy or magical transformation. This alchemy offers a fluidity of perception usually hidden to us when we are living in 'ordinary reality', a different way of experiencing presence, place, time and movement.

I use my work, 'Coat', as a case study through which to describe this way of working. I name the shamanic practices I work with that support maintaining embodiment alongside a shift into a shamanic state of consciousness allowing me to fully enter the work and non-ordinary reality.

"My friend they are returning again, all over the earth, ancient teachings of the earth, ancient songs of the earth, they are returning again upon the earth."

From words spoken by Crazy Horse, Lakota, Sioux. 1840-1877.

The hunter prepares for the hunt with ceremony and with prayer.
She is utterly still, she brings her attention to her breath, she is centred in her body, her awareness is active in all directions, she is filled with the present moment.
The hunter is awake to all that is.
The hunter is tracking with all her senses, she becomes aware of the presence of animal, she has 'seen' the animal with all of herself.
The hunter asks permission of the spirits and then respectfully begins to enter and merge with the field of the animal. She asks for the gift of the life of the animal, that she may receive the life-giving nourishment of its body.
Within the stillness the animal consents to give herself away.
The hunter is centred in her body, her heart is open and connected to the heart of the animal.
With a prayer she releases the arrow from her bow, its tail trembles in the air as the tip travels towards the heart of the deer.

The tip of the arrow enters the heart of the doe, at the moment that she willingly gives of herself.
The hunter gives thanks to the spirits and receives with gratitude the nourishment and life of the doe.
Her blood, her flesh, her skin and bones will serve the hunter and the people. The hunter and the people will eat the flesh of the doe with gratitude, her skin will become medicine drum. The song of the doe will sing on and on through the heart beat of the drum.

It is said that in the days of the 'old ways', in the times when humans more consciously relied upon the earth day to day, for food, shelter, clothing and tools, that ways of being were born out of and through an instinctual inhabiting of both the body and heightened awareness of the movement within the cycles of the earth and the sun and the moon and the stars. In these times, humans were seen as 'living at the bottom of the sky ocean', an indivisible part of the biosphere of the earth.

These were shamanic teachings. Shaman is a word from the Tungus people of Siberia and translates roughly as 'the one who sees'. The shaman translated messages from 'our relations', winds, moon, bodies of water, animals and plants, for those who did not understand them. The shaman once lived on the edges of community; she was the healer, she knew where the medicine plants grew and when to hunt for animals for food and could divine answers to questions from the people. The primary work of the shaman was to maintain balance, balance within the self, the other, the community and the earth; she followed the path of wholeness and balance in the ways she knew and that had been given to her by those who had gone before. She was a messenger between the worlds which she always approached, as a close relation, with great love and respect.

Once again, many humans are seeking wholeness and balance, within ourselves, our communities and the greater cosmos.

The contemporary growth in the study of a broad base of shamanic teachings, which long ago emerged in similar form through many and different cultures and places on the earth, sits at the root of this call to understand how the ancient teachings may be reawakened and incorporated into all aspects of life in the 21st century. It is my seeing that there is a renewed interest in this ancient understanding as an answer to a call to remember at our core who we humans are. This call is arising as we become more aware of a growing state of imbalance within ourselves and within the whole biosphere of the earth.

For well over a decade, I have studied and practised both shamanic teachings and Earth Wisdom teachings through Northern Drum Shamanic Centre and environmental movement supported by the work of Helen Poynor. As I have become more familiar with these practices, I see that my physical body, my sensory body, my emotional body, my creative impulse and spirit align with many of the ancient teachings that I have received. So much so that I now live and work through these practices absorbed into my daily life and they shift me into a new sense of living. In this way I apply shamanic practices to support all aspects of my work and life, environmental movement practice, performance, ceremony, healing work and as guardian of Vision Quest. Embodiment of the teachings connects me to their ancient lineage; this sense of connection allows me to trust the responses of my body and to respect and value the information received through all internal systems, and from the elements and all living beings that I exist within and as a part of.

As I enter practice, I call the nature and sensibility of the ancient hunter to become awake within my body and through a wider awareness of time and space. I am informed by this awareness. It allows me to work from within a deep level of intuition on all planes. Within the present, I am opening towards and following at the same time. As I follow I am able to remain connected moment to moment to a knowing and understanding of the truth of what emerges. I have a sense that I am following an ancient thread of creative impulse that exists and at times touches into my field well before I have any conscious sense of it. To begin with, when the thread begins to come into the physical realms, I may have no conscious idea of what it is attached to. I and my way of working become the portal or entry point for the creative impulse to emerge into the physical. I simply align with it and allow my own creative spark and imagination to join or to follow the flow of its emergence.

Since before I can remember, I have been curious about the nature of things transformed or left behind; pebbles, owl pellets, bones, feathers, driftwood and much else. Soon after beginning the active study of shamanic practices I began to find and collect bones, vertebrae in particular, with a curious earnest. As far as I was aware at the time, I was collecting with no particular intention other than curiosity and the joy of finding. I seemed to come across animal vertebrae wherever I walked. In time, I had accumulated a large collection of bones of both birds and mammals both from the ocean and from many lands. One day, as I contemplated the bones it occurred to me to string the animal bones into one long spine in mammalian structural order from neck to tip of tail, to see what would arise. What arose next was a ripple and a sense of magic, like an old tale come to life and then almost at once I received a strong image of wearing a coat, akin to a tail coat, made of blank painter's canvas, with a tail long enough to trail upon the earth behind me as I moved. The spine of animal bones would be attached to the neck of the coat and hang free down the back. The blank canvas of the coat and the attached spine, could, I perceived, serve as a conduit for the emergence of ancient stories. At once a call was made to the ancient hidden tales of the earth and inhabitants to be remembered and awoken through the gentle brush of the tail of the coat, to become painted into the warp and the weft of the fabric and thus through their essence return to the present.

Body and Awareness

Fig. 1: *Coat. Hilary Kneale. Omey, Connemara.* Photo: Christian Kipp

In this chapter, I look into characteristics of embodiment and shamanic sensibility in relation to body-based creative work, through the case study of one of my own works. I contemplate 'Coat', as the core study for this chapter; it is a piece which has developed over a period of years, emerging as a physical presence inhabiting different places on the earth, as performance, as story and as film[1].

'Coat' is one strand in the tapestry of my work and experience that calls in a shamanic sensibility. I have chosen it as a vehicle for my inquiry because the strong narrative of the work holds clear examples of what I refer to in the title as 'The Alchemical Body' and it illuminates three key shifts within body, awareness and sensibility. These are:

- Cultivating an embodied shamanic mode of attention.

- Entering the portal. The point of shape shift.

- A wider sense of movement and time. Entering the 'Mystery'.

[1] www.hilarykneale.com

Cultivating an embodied shamanic mode of attention

Pulling on the coat feels like an ancient skin regrowing. I sense my arms inside the sleeves, my hands emerge awake and vital. I see hand, skin and bones and claws and paws of all potential. The spine of animal bones hangs against my back, presses against my own spine through the fabric of the coat. As I move, there's a rattle of a song between spine and spine. I crouch by the stream at the place where two clear springs tumble from under rocks in a bank beneath an ash tree, the water tumbling and merging into the greater moving body of stream water, water coming together with other water. I sit quiet. I listen wide.

I begin to move slowly, easing my way through the thickening air of early dusk; it is touched by soft golden moonlight, the air shimmers around me in the softening light. The tail of the coat slides through the stream as we cross and re-cross her path, cold clear waters shine out under the moon, painting the moonlight over the tail of the coat. The fabric of the coat glistens, its warp and weft newly vibrant, like sea washed pebbles left by the outgoing tide. The colours of the coat's many travels are now a memory within the tones and hues of the canvas.

I and the coat begin to merge. Touching the earth footfall by footfall we move away from the stream, although her presence travels on with us as song in the air. We pass over thick moss-covered rocks, moving four pawed then rolling beneath the reaches of the great beech, itself winter bare and cloaked in soft moss. We sit as one, taking time to enter deeper into the stillness of the present moment. What is the sound of the story without words? We listen to the whisps of what has gone before. We are awake.

The shaman traditionally 'journeys to other realms' to the sound of the drum, in meditation, and through ceremony to seek information in relation to wholeness and balance. Sometimes she may 'shape shift' or merge with another creature to assist with the task at hand. In my experience, a shamanic state of consciousness creates a connection supported by and beyond the imagination and through into the greater mystery of all existence.

I am wearing the coat, preparing to journey into its hidden realms. I allow my breath to deepen into the lower energy centres of my Base and

Will centres[2]. My attention is fully in my body, heart open, back awake, creating a three-dimensional awareness of my physical being, a sense of renewed vitality courses through my body. I sit in a state of stillness, eyes open in soft gaze, adopting a wide peripheral vision. Calling support and protection from spirit, I breathe deeply into all my energy centres to allow them to open and balance, I am connected deep into the earth, through my Base centre. When I feel ready, I sense a strong support and protection from spirit, which comes into my awareness as a thickening and a quickening of energy around the container of my physical body and also a sense of a supportive or 'knowing' presence. With the power of my intention I enter a shamanic state of consciousness and wider awareness and merge with the coat. I merge with the soul of the coat, the animus, and shift shape into our dual nature, spirit and matter join together. I sense the merging though a subtle sensory experience and quality of body-mind and expanded time and thence begin to move. Slowly at first, my fingers and hands and then my arms fill with the energy of life force, like the power of rising sap, moving out from the animal heart of myself; I follow this impulse: now I am both of the earth and with the earth. This quality of movement feels akin to the first power of creation, unknowable and yet somehow clearly mapping itself into repeating shapes and cycles as it emerges from the core of existence.

Entering the portal. The point of shape shift.

We are awake.

I tremble within the shifting states of the portal. To begin with I have an awareness of the call from both sides, my focus is in flight as though pulled and honed by the tip of the arrow travelling before me. I, as Coat, am held within the portal of its slipstream. All around me the colours of plants and trees appear to be growing more vibrant,

---

[2] Energy centres of the body: centres of energy are seen both as points, and as existing around the body. They often need some attention in order to come fully into balance and to be charged with energy or life force. The base centre is below the genitals and the will centre is below the navel. They are the first two energy centres that help us to feel grounded in the physical body and are the source of creativity. They are seen as key to a sense of safety and enjoyment of living in the body. A detailed account of all the centres can be found in (Luttichaü, 2017).

*everything shows itself in great detail. What I see imprints itself upon
my retina, details are slipping into me under my mind and entering
my memory to become a part of my animal body, a part of the place
and time of my existence. I become Coat, I tingle with excitement
through the shift of shape, bones are a-rattling. I call the layers of
unknown deeper in, diving into the depths of perception. I am filled
with wonder and the curiosity of experiencing everything in its purity,
as though a young child.*

*I am crouching low, low to the earth, tail behind me as though ready
to pounce or to run or to fly.*

*In a lifetime or in a moment I sense that I have passed through the
portal.*

The ways of passing through the 'portal', can be unpredictable and
unknowable. In this instance, before my call can be returned by its echo, I
sense a need to crouch low to the earth, limbs animal wise, aware of all
sensations in the structures of my feet and hands, now rising through my
spine against spine. Stories begin to jostle and sing in the air around me
and under me the earth is whispering.

I perceive that as I enter a shamanic state of consciousness and so move
through the 'portal', 'Coat' and I surrender into the unified field of energy
in time and space, inside which I become fully present in body, mind and
spirit. The shift into another state of being is palpable, I am 'other', I have
entered another 'realm', at the same time I am aware of my surroundings
and a sense that I have come home to myself more fully. Having passed
through the 'portal', nothing exists other than an expanded and timeless
present moment, held within the movement of all that has passed and all
future potential. I sense that the life force or energy behind everything
becomes 'visible', from the moment of shape shift until I re-emerge
through the 'portal', at which point I am simply aware that I have returned
to 'ordinary reality' and my ordinary self.

Entering the 'Mystery'

To attempt to unpick experiences that only bloom for a moment is perhaps
to destroy them. I am clear that I cannot *think about* being of the 'Mystery'
and *be* of the 'Mystery' at the same time; one potentially dissolves the

other. At the same time, I see that words speaking through the remaining imprint of experience within the body-mind have the potential to return through the portal and with a quality of echo, speak from the essence of 'Mystery'.

I am speaking with Chris Luttichaü, founder of Northern Drum Shamanic Centre, author of *Calling us Home*, in which he writes about the value and relevance for our times of shamanic teachings and practices. We speak about the nature and quality of the 'Alchemical Magician' or Alchemical Body, which is referred to in the old teachings. A shifting of shape based on oneness, a movement of both perception and body is the energy that shamans call upon to transform themselves into something other for purposes of seeing, teaching and healing. As Chris speaks of the 'Alchemical Magician' he describes it as a 'core shamanic term' saying that its meaning is akin to the alchemical process (Luttichaü, 2017: 143). The Alchemical Body is seen as embodied transformation, bringing a capacity, in shamanic terms, to shift from 'Ordinary Reality' into 'Animated Reality'. He describes 'Animated Reality' as a oneness, non-duality and interrelatedness and says that through inhabiting this quality we can fully enter the circle of life, the sacred hoop, the 'Mystery'.

The 'Alchemical Body' holds the capacity to bring change through movement of mind, body or shape, to change from moment to moment like a little child and with that comes a capacity to renew vitality and to live with integrity.

David Abram, cultural ecologist and environmental philosopher, writes of his own experience in relation to shapeshifting. He describes the moment when, after many months of study and practice with a shaman in the Himalayas and after many failed attempts, he is finally able to shape shift into raven. Through merging with the bird, Abram experiences himself flying through the high mountain valleys of Nepal.

> A vertigo rises from my belly into my throat and I'm falling, I'm falling. Gonna die for sure... I'm balancing, floating, utterly at ease in the blue air. As though we're not moving but held, gentle and fast in the cupped hands of the sky. Stillness. Through a tangle of terrors I catch a first sense of the sheer joy that is flight.
> (Abram, 2010: 256)

Stillness of mind is as much a key to inhabiting a shamanic state of consciousness and entering the 'Mystery', as is movement. Stillness of mind or 'body-mind' supports wider connections to the macro and the micro within all bodily systems bringing increased vitality and potential for movement. Movement does not replace stillness, the two states become one. I refer to movement in its widest sense, movement as presence in the body, as cycles within the systems of the body, and all aspects of time, and within the collective consciousness of the earth and the wider cosmos.

In a lifetime or in a moment I sense that I have passed through the portal.

Birds songs sing out the last light of day. Grandmother moon rising at the full point of her monthly flight. Night sounds herald the evening star.

Hush closes round the sounds hush hush hush little one, sleep now, they are waking.

I enter Coat, we become. Again there's a pulling body to body.

Deep in the belly is a hearing of them waking, hearing the old ones beginning their nightly journeys. It is a low rumble of a sound that shimmers the sky to speak of thunder. A rubbing of face and paws along the fox path to seek the way through the tangled web of the growing. Body remembers where the light falls at times when grandmother moon shines through the shimmering darkness. This night moon hides her fulsome form cloaked in cloud. Animal rememberings find the way as pores open up to them. Lower and lower to the earth we tumble, until we smell our way through to the night scent of oil beetle. Eyes drink in glistens of rain caught as diamonds round the openings in earth bound spider web. She has woven time after time and time after time the great weaver repeats herself. Arching backwards and pouring over the mighty limbs of fallen trees, we descend into the earth, through years of the slowing we enter her body.

After time and so so quietly a song begins to sing, to open lungs and heart and throat to sing a name and to sing the earth and the moving water and the falling trees. Teeth gnawing the edges of the earth where it is opened by the stream to depth of bone. Entering the stream

to become a glisten of a tail in water under the moon, of moon. Lumps
of silence easing out of the earth become song. Knots of words held
under the fallen trees emerge tendril like over fingers and claws. Eons
of story begin a rattle through the spine of animals rolling and
cajoling amongst the night dark bluebells. Songs become a
remembering once more. Limbs flail wildly as words pass as stones
through liver and kidneys and spleen over and over they crawl out
through the groaning places. Stories are leaving the warp and the weft
of coat, wildly sparking like fireflies as they meet the night air.

Settling all about, we see that they are stories returned from the dark
sky.

We are of the dark sky.

We lay amongst them in the arms of all, lay amongst them resting in
the arms of all, hear all…

It is through awakening to this quality of movement and with a connection
into expanded time that I am now called to enter the 'Mystery'. In the
nature of the Alchemical Magician, my shape shifts, I become 'Coat'. I have
a sense of being connected to and aware of everything, all is in motion, all
is still. Now is the time to dream the way of the future.

She stands as deer in the quiet of herself. She will give of all of herself
when the moment comes. Time expands, the hunter, the arrow and
the doe are held within a slipstream of time, meeting at the point of
the arrow.

References

Abram, D. (2010) *Becoming Animal: An Earthly Cosmology*. Pantheon Books

Luttichaü, C. (2017) *Calling Us Home*. Head of Zeus

Further Reading

Eaton, E. (1982) *The Shaman and the Medicine Wheel*. Quest Books

Gintis, B. (2007) *Engaging the Movement of Life: Exploring Health and Embodiment Through Osteopathy and Continuum*. North Atlantic Books

Halifax, J. (1979) *Shamanic Voices: A Survey of Visionary Narratives*. Dutton

Harner, M. (1980) *The Way of the Shaman*. Harper

Hay, D. (2000) *My Body, The Buddhist*. Wesleyan University Press

Turner, K.B. (2016) *Sky Shamans of Mongolia: Meetings with Remarkable Healers*. North Atlantic Books

---

Hilary Kneale is an independent interdisciplinary artist, who readily works in collaboration with others from different fields. She is a published writer, educator, guardian of Vision Quest, movement practitioner and healer, living within her own quest to remember the true nature of interrelatedness.

Through her many ways of working, Hilary calls strongly to the ancient stories held deep within the earth and listens to the return with all of herself. Having trained over many years to embody, to develop and to teach practices with support of the work of Helen Poynor, and Northern Drum Shamanic Centre, she holds ways of opening the body, heart and mind, that can re-awaken the native soul.

hilarykneale@gmail.com ~ www.hilarykneale.com

The Changing Body

Awareness is Alive

Paula Kramer

Abstract

This chapter considers and discusses awareness as alive and changing, constituted between body-mind, environment and movement – a kaleidoscopic, shifting, living and never fully controllable potential. Whatever we do, our awareness constantly changes – it is inclusive, and can include more; it is expansive, and can expand more.

Rooted in movement and performance practice, two entwined proposals are formulated here that support the capacities of our awareness. One is the commitment to bodily practice: regular consideration, preparation and training. The other is more of an inner gesture of inviting, or calling, or thinking possible, that awareness *can* change, expand and become more acute. Whilst in the end beyond our (total) control and never twice the same, becoming acquainted with and working with awareness offers the possibility of consciously experiencing awareness as expanding beyond the self, including others as well as the world at large, seeping also beyond moments of practice and performance into daily life.

I lie, folded, on one side, my body touching rock. Wind and air and sound touch me also. Here a small universe opens for a moment, cradled in the sea. The rock feels like the beginning and end of everything – a crag, surrounded by water.

Deep time, time of rocks, thousands of millions of years. Head inland, feet towards the sea. I hear the waves, the water, all sounds. I can release into this feeling of lying and listening, all alone, on a small island of rock, somewhere surrounded by water. I find solace in sensing my own smallness and this way of being carried, being held, on the face of the earth.

I can feel my body, soft, having form. My fingers touching rock, touching air. There are winds, waves, birds, boats, light, planes, humans, rocks, shrubs, grasses, trees, sky. But there is also just my body, just this form, just this touching through skin, supported by rock, cradled by the wide, open sea.

(Recollection of the beginning of 'On the surface of time', performance, August 2018, Suomenlinna, Helsinki)

Here I begin

'Awareness' in this text is understood and discussed as a faculty of the human body-perception that is alive and changing. It is inclusive, and can include more, it is expansive, and can expand more. In some way this changing awareness I speak about here can feel like tuning from mono to surround sound. The world becomes more (alive), colours more (vibrant), details more (available), sound more (differentiated), scales and proportions more (clear), three-dimensionality more (tangible). And if our practice serves us well in this particular moment, we can develop and follow our moving-dancing in a resonant, relational integration into this world and feel at ease[1].

---

[1] The combined word 'moving-dancing' is frequently used in the context of *Amerta Movement*, the movement practice which nurtures most of the ground on which my own work and practice grows. Reeve translates Suprapto Suryodarmo's term *Joged Amerta* (a specification of what is more broadly referred to as *Amerta Movement*) as "the moving-dancing nectar of life" (Reeve, 2010: 189). I've heard the word often in movement workshops led by Prapto and frequently use it in my own writing and teaching. To me the hyphenated term contains and speaks both to daily life movements and to crafted, emplaced, embodied dance, both of which are formative for *Amerta*.

I recognise a difficulty in the use of the term 'awareness' that I follow here, which can lead to a lack of differentiation of awareness from say perception, consciousness or feeling. I counterbalance what might seem blurry (from the perspective of neuroscience or cognitive psychology) with detailed and concrete descriptions that draw on movement practice, which I hope will be useful anchors and points of reference for movement practitioners and performers, particularly those who work outdoors.

As an overarching theme, my work as a movement artist and artist researcher focuses on intermateriality. Through embodied and movement-based artistic research practices I have been, and continue, researching how things, objects and materials of different orders interrelate (and not), specifically in the context of outdoor movement, performance practice and choreography. Awareness plays a relevant part in this, because it becomes, manifests and changes in the relationship *between* mover and environment, materials and atmospheres, rather than independently. It is shaped through our moving-dancing and deeply interwoven and entangled with movement and movement practice, as well as with our material and immaterial surroundings and contexts.

Philosopher Alva Noë, who engages deeply with perception, consciousness and the human body, makes similar remarks in relationship to perception in his book *Action in Perception* (2004). The book as a whole elaborates the proposals that "*[w]hat we perceive* is determined by *what we do...*" (Noë, 2004: 1, original emphasis); that perception is "...not a process in the brain, but a kind of skilful activity on the part of the *animal as a whole*" (ibid: 2, my emphasis); and more specifically, that "...perception is in part constituted by our possession and exercise of bodily skills..." (ibid: 25). Whilst my writing has a different emphasis from his, draws on a different field and answers different questions, I do shape a related line of argument in relationship to awareness, bodily/movement practice and environment/surrounding here. Our awareness unfolds and becomes present in relationship to how and where we move – i.e. our bodily skills and "what we do where", "what our bodies do where".

The movement practice on which this writing draws is heavily informed by the wider context of *Joged Amerta/Amerta Movement,* a movement practice that has been developed since the early 1980s by Indonesian movement artist Suprapto Suryodarmo (Prapto) in a unique intercultural setting between Java, Indonesia, where Prapto is based, Western Europe, where many of his early collaborators and students are based, as well as North America, Mexico, Japan, India, Korea, the Philippines and

Australia.[2] I have personally experienced *Amerta* primarily as a movement practice in the context of contemporary experimental dance practices, rather than as self-exploration, therapy or spirituality, which are all strands that interlace in the practice of *Amerta*.

In my experience, *Amerta* is a highly complex and timely movement practice that supports movers in learning to practise movement 'in relationship'. This relationship includes oneself, one's environment and other movers – all of which might pass by or drop into our awareness. To practise *Amerta* means to cultivate a highly differentiated ability of sensing and following – in movement, but also in association, imagining, dreaming. And at the same time forming and choreographing movement as it emerges. Whilst we practise these skills, we continue to notice that we are embedded in a wider context that includes inner and outer conditions, colours, weather, sounds, songs, tasks, people, plants, atmospheres, intentions, animals, textures, histories, materials, structures – and so on. To practise movement in this way, i.e. 'in relationship', 'in context', is key to how I understand, make sense of and work with awareness, intermateriality and *Amerta* more generally and is a continuous, resonant part of the movement experiences I draw upon throughout this text.

Change and Expansion

Using the support of words to delineate, I focus here in particular on two aspects of awareness, which highlight the ways in which it can be considered to be alive as well as intermaterial and constituted between body-mind, environment and movement. One is the capacity of awareness to *change;* the other is the capacity of awareness to *expand.* In my understanding, the aspect of change is the more general notion here, whereas expansion is one particular example of such a change, which includes both a widening focus and a heightened attention to details. Expansion thus not only means 'more' and 'wider' here, but also an expanded tuning into detail, into intensity.

The notion of awareness as *changing* locates and understands awareness not as a fixed or set faculty, but as a potential of our bodies and brains that

---

[2] This text was written prior to Prapto's death in December 2019. References to him and his work are therefore written in the present tense. Whilst there was time to make the grammatical change, I decided not to. Maybe this choice can contribute to the vision, hope and possibility that in practice and in the further development of *Amerta,* Prapto's work and way of working will remain exactly that: present.

is highly malleable, tied to a myriad internal and external factors and, in this sense, living, alive. I draw this sense of aliveness from basic understandings within the genealogy of *Amerta*, in which also space is conceptualised as "alive and happening" – an aspect passed on to me especially through working with Bettina Mainz. I would argue that it is one of *Amerta*'s specificities to consider the world that surrounds us – as well as any moment in time – as alive, as living, as happening. In the same way, awareness is alive and happening, is changing and interdependent with what we do, how we tune to our bodies and our movement and what is around us.

One of the most significant and easy-to-notice changes in awareness is, in my experience, expansion. There can be, through movement practice, a significant, noticeable shift in awareness – from an awareness that might be experienced as narrow, limited, fragmented, tired or restless to an awareness that is expansive (sometimes surprisingly), that has a broadened radius and that can include also unforeseen aspects, dimensions or details.

Awareness can become at once (as well as incrementally) broader as well as more detailed. I might simultaneously take note of the bird overhead, the noise of an engine from afar, the smell of the spring, the way I breathe or how I am positioned and scaled in relationship to other aspects present on a site.

> As I lie down in the grass I can work with the sky moving. My back in the grass, my hair getting wet, I move sideways with the clouds … there is a lightness I can draw from, a distance, a cloud texture, …shades of grey moving, for a moment the rain has stopped, the air is cool against my wet face.
>
> (movement diary, Stoneleigh, UK, 25.04.2012)

Body as base and the relevance of inner gestures

Two preparatory moves support the unfolding of awareness, two moves can make space for change and noticing change, as well as for awareness to expand, and to become more acute, detailed and inclusive in this expansion. One such move is a commitment to bodily practice, the regular consideration, preparation and training of our bodies and their capacities. The other is more of an inner gesture of inviting, or calling, or thinking possible, that awareness *can* change, expand and become more acute.

When turning to the first preparatory move, that of bodily training and practice, I refer to dance and movement practice, staying within my chosen frame of reference. From here one might extrapolate to other bodily practices such as hiking, climbing or also meditation and mindfulness practices.

Importantly, practice does not mean to rehearse or inscribe a standardised reaction of our perceptual system that we can then retrieve. But practice allows us to become first acquainted and then familiar with how our awareness is related to, and forms part of, movement practice. Awareness is thus not an independent faculty, but develops and changes as an integral part of what and how we practise. We begin to recognise and differentiate shades and layers of a changing awareness and we learn not to despair if awareness does not 'pop open' as we may have expected. We become familiar with different ranges of expansion and we develop a frame of reference for what it feels like when our awareness begins to offer also, as an example, a place-in-relationship, a place-in-context, for ourselves in this plentiful world.

With the second preparatory move, an inner gesture of inviting or calling into presence the possibility of a changed awareness, I refer to the gesture of adopting a changed perspective within our mental frameworks and ontologies. What if I assume that my awareness can change continuously? What if I allow myself to experience my awareness in interdependence with my body, my practice, the world? What if my awareness could open and include dimensions and aspects that I have not noticed before? We can formulate these possibilities both internally or hear words, sounds, song or chanting from a teacher, mentor or peer, which stimulate such changes. Someone might explicitly encourage us to open our awareness to what surrounds us. In other cases it is the generation of a specific atmosphere that supports us in opening to sounds and depth of field, to notice what we notice, see what we see, sensing ourselves in relationship to how and where we are now.

When Prapto teaches, his voice (and a drum) often quite continuously form part of his facilitation. Through song, chanting or speaking to the group or to a particular mover, he accompanies, supports and tunes into an individual, a group, an atmosphere or a situation. Similarly I know this from working with Bettina Mainz or now teaching people myself, that an added sound, hum, noise or song might shift the atmosphere in a site (or in a studio), which in turn can also bring other aspects (beyond the sonic) – to the awareness of the movers. These might be highly varied and can include colours, shapes or smells, constellations in the space, relationships or choreographic forms, bodily sensations or memories and associations.

Rather than existing as separate faculties, body, bodily practice, invitation and calling into presence are intensively entwined in their relationship to awareness. They interweave also in our dancing and operate on a broad and changing spectrum. Through the route my training as a movement artist has taken, I would say that anchoring well in my physical body is what has been established in me as a kind of baseline commitment in my practice. Whilst I always also tune my inner attitudes, such as actively or through habit assuming an inner position that invites and allows my awareness to change, I do this as part of my physical practice.

Especially in the context of performing for an audience, I rely profoundly on 'staying close to my body'. Such anchoring in the body then offers a physical base in relationship to which awareness can change, expand and/or become more acute. Without a sense of body, there might be a sense of expanded and/or more detailed awareness, but it can easily be lost as soon as the body begins moving. Or, awareness might already be so 'far out', that it becomes challenging for the mover to even begin moving. Debilitating questions such as 'why move at all?' or 'what am I doing here?' might arise and obstruct the unfolding of movement. What I have found to be practised deeply in the wider field of *Amerta* is the cultivation of an *expanded awareness in movement*, which requires clarity in one's own embodiment as a base.

The extract from my movement diary prior to this section speaks of a bodily involvement and physicality from which a broadened and more detailed awareness expands – "as I lie down… I can work with the sky moving". Physical and sensorial details are clear and available to me – I can sense how I lie (my back in the grass), how I move (sideways), what is cool (the air) and what is wet (my face). From here I was able to connect with, and draw from, qualities I sensed in the sky (a lightness) and the clouds were noticeable to me, in that moment, in their specific colouring and texture. Such broad and fine noticing does not happen as a given and I experience such moments as especially satisfying – a changed and expanded awareness is coupled here with a specific movement that emerges without having been planned.

Expansion and Detail

A significant shift in *what I notice* is often the first recognition within myself that my awareness is changing. A shift takes place, sometimes sudden, like a switch being flipped, sometimes more slowly and gradually, in widening circles, until I notice that I am noticing *differently* and/or more. My awareness might rather generally open, widen, *expand* until I come to realise that I have an immensely broadened experience of the site. This might include that my ability to hear becomes more nuanced as well as broader (noticing sounds from a distance, for example, or hearing a sound I did not hear before) or experiencing a kind of bird's eye view of the site of practice. Likewise, simultaneously or separately, previously unnoticed *details* of a site might appear to me (the weeds over there, the shadow on the wall, the reflection in a window, the colour of the grass, the shape of a tree) and become foregrounded in my awareness, even though I might have looked at these aspects many times before.

I have experienced such changes in awareness both directly upon arrival on site as well as during and through movement practice. From my current understanding I would argue that the first opening ('upon arrival') speaks of a kind of first-instance shift, a kind of first tuning into site and entering into a phase of practice. It is, maybe, a signalling of a first layer of body-mind readiness: now I am here, now I am noticing that I am here and I begin to practise already in these first steps on site, during the procedure of, quite literally, 'arriving'. The change of season, the specific light or weather conditions of today, a vehicle that was not here last time, an aspect of a plant, a previously unnoticed texture or structure, the perceived height of the sky today, the number of birds circling above or not – with and through these first aspects that come to my awareness I arrive and step into a site. Both an expansion of my awareness as well as a noticing of more details can take place upon arrival, as a first tuning.

Such a first-instance shift can then deepen and expand through movement practice and a site might burst into details through changing my bodily position, through leaving the habitual, human upright position and meeting a site in and through a body that becomes moving form and material, an interface with the world through dancerly expression and dialogue.

> *...as I lie down I begin to see movement everywhere. The clouds begin to move, the reeds in the wind, the water, the grass, the birds –* **everything**. *As I was walking the world was still – apart from the wind – I mostly perceived the quietness of the rock. Both is here, always, it's a matter of what I tune into.*
>
> (movement diary, Suomenlinna, 21.04.2018)

With a changed bodily position (for example from walking through lying down as described above) and beginning to move/practise from the sense of body-site contact, the aspects, quantity and/or quality of what falls into my awareness can change significantly. It is important in the context of movement practice and/or performance preparation then not to be swept away by the sensorial input, but to simultaneously stay present in the body, not getting lost or overwhelmed by all that is offered to my/our 'noticing'. The task at hand is to interweave the bliss of an expanded and heightened awareness with a kinaesthetic awareness of how we are positioned, what our bodily form is in relationship to other forms, to support our movement practice and allow for a sense of following rather than planning movement.

> *The powdery quality of the snow. the colourful surface of the rock underneath. the tiny, tiny snowflakes. the "evening" (afternoon) adding blue to everything. The layered clouds (some blue there, also) – grey, grey, white, grey, light blue, and so on. the voices of the two women walking their dogs. …I see, feel, hear, taste, smell, sense, winter all around me with its quietness and cold that later produces deep exhaustion as my skin warms up again, feeling dry and hot, inside now.*
>
> (movement diary, Suomenlinna, 11.02.2018)

This example from Suomenlinna speaks both of a broadened and expanded awareness, as well as an acute sense of detail and nuance. It was a moment of practice in deep winter, with notes taken after moving when I was back inside. I notice the textural quality of snow, the size of snowflakes, the colour of rock and snow. I experience winter through multiples senses (see, feel, hear, taste, smell) but my awareness also clocks the larger whole, including the light of the afternoon, that looks like the evening in the Nordic winter, the layered clouds far above me and voices coming from afar.

This sense of a widened, expanded awareness can also be much more explicit, foregrounding itself or dominating my experience of my awareness. My whole perspective on how I am positioned in relationship to a site can change or tip completely, from a perspective of looking from within myself into the environment to a sense of being able to sense and see myself from a point of view much further afield. Another excerpt from my Suomenlinna notes describes such an experience:

The skin, the crust, yes – the below, the above, the earth, yes, the earth on me and me on the earth, yes and the rock feels so small, suddenly.

(movement diary, Suomenlinna, 04.09.2017)

What is captured here is a shift of feeling myself, my size and relationship to the rock, from within me and then zooming out, seeing myself tiny on the rock and even further, sensing this assemblage of rocks, this small island, as a minuscule structure on the planet as a whole, part of an ever expanding universe.

...this incredible clarity or force? Or what is this – that the Federal Ministry of Finance sends from behind, I mean purely through its architecture – in this light, today, incredible – beautiful in this way 'it hits me', it is present, so mighty and manifest, and invites me to also be like that.

(movement diary, Martin-Gropius-Bau, Berlin, 24.11.2016)

An example that speaks more exclusively to the noticing of details stems from practising in Berlin, on the south side of Martin-Gropius-Bau. The clear architectural structure of the Federal Ministry of Finance, which I could see diagonally in the distance from the site of practice, hit me with its overly clear architectonic structure, its many lines and corners, windows (fieldnote quote above). This building really is monumental in the city space, yet I had not paid much attention to it on my previous visits to this site (how come I did not notice this dominant building? I later ask myself). It directly affects my practice on this day, clarifying, stabilising and strengthening my moving.

Writing about this does not come without ambivalence – this building is one of the most monumental Nazi structures still left from the visions of founding a world capital 'Germania'. It is an imposing, much too large building in the wider cityscape, very stark and grey, looking partially like a prison. As such it is not a building that I would particularly resonate with if I were to pass it in daily life; much on the contrary. But on this day it fell into my awareness differently and its clear form in the space offered a flavour of its structure to my dancing, affecting my movement.

This moment (or a phase before or after) most probably also included a widened perspective (the building for example is located across the street at some distance, so not right in front of where I was). Yet what I still sense mostly in my memory of this experience and in my notes is the appearance of a specific aspect or detail of the site in my awareness and the palpable effect it had on my movement.

Similarly, and on another day, a tree that I had been around several times suddenly attained significance and appeared to me differently:

The birch I see like for the very first time, the white trunk, the white branches, the white snow. the roots appear as clear forms even under the snow.
(movement diary, Martin-Gropius-Bau, 31.01.2017)

On another day it is the light that falls into my awareness. I add this here to emphasise that not only material structures 'make' or 'compose' a site, but also atmospheres, weather, seasons or light impact upon practice and can become foregrounded in our awareness in particular instances.

The sun made a dramatic difference today. adding shadows, three dimensionality and texture. I could dance here today also because of the light, because of the space it offered, it created...
(movement diary, Martin-Gropius-Bau, 26.01.2017)

I recognise such noticing and being affected by aspects and qualities of the site as a familiar part of becoming acquainted with a site. On each new visit I become aware of another aspect, another detail, another angle, another quality that I have not noticed before. Each day offers itself completely differently, with different weather, atmosphere and activities/life/happenings at the site. Once such aspects have entered my awareness and formed part of my experience of practice, they become anchors that I can return to in later instances of practice and especially when moving towards building a performance score.

Concluding thoughts

Importantly, none of the aspects I discuss in this text can be controlled, fixed, pinned down or repeated twice exactly the same. It is both magnificent to experience how broadly awareness can expand and what it is capable of including, as well as excruciating to learn that awareness cannot be controlled (though practice does make a significant difference). Thorough practice supports our ability to accept that sometimes awareness expands and excites and sometimes it remains narrow and circles around whatever troubles and impossibilities might arise. To learn to accept this without categorising a less-than-expanded awareness as failure is probably one of the most relevant steps in coming to terms with the volatility and aliveness of awareness in movement practices such as *Amerta* and ultimately in life more generally.

I would like to highlight in closing a consideration of awareness first and foremost as a kaleidoscopic, shifting, living and therefore not controllable potential of our bodies, brains and sensory-perceptual capacities. I suggest thinking of awareness *through* practice and allowing several components to be present – awareness might be shaped and affected by very physical and material aspects, by manifest details, structures or colours, and simultaneously the weather, an atmosphere, an angle of light, as well as our speculative or intuitive associations, affect our awareness. It can already change as a marker of arriving on site, with another layer opening through movement practice. Or we begin with bodily practice, however we are, and our awareness expands and/or becomes more acute at a much later point (or never, in relationship to this moment).

Whilst awareness in movement practice is neither solely this nor that, it is without doubt a powerful faculty of our human existence with extraordinary capacities worth engaging with as a movement practitioner. Paying attention to, becoming acquainted with and consciously working with awareness offers movement practitioners deeply relevant experiences that expand beyond the self and can include others as well as the world at large, seeping also beyond moments of practice and performance into daily life.

Acknowledgements and Lineage

I first came across *Amerta Movement* through training with dancer-choreographer Bettina Mainz (beginning in the late 1990s) and am still touched by the experiences of these 'early years'. Significantly later I completed two years of Helen Poynor's *Walk of Life training programme in Non-stylised and Environmental Movement*, a context strongly influenced by *Amerta* as well as the work of Anna Halprin. I worked with Prapto himself mainly during the last decade of his life and deeply miss his voice, presence and movements that brought so much to so many. Thank you Prapto, for what you brought to movement, to life.

For central points of this text, such as the relevance of 'body as base', I wish to thank Helen Poynor. I greatly appreciate the way she makes her sustained practice available to others – allowing for the development of insights in relationship to specific experiences, as well as nurturing one's own, individual approach and understanding of movement practice more generally.

References

Kramer, Paula (2012): 'Dancing in Nature Space – Attending to Materials' in S. Ravn and L. Rouhiainen (eds.) *Dance Spaces. Practices of Movement.* University Press of Southern Denmark, pp. 161-174

_____ (2015) 'Dancing Materiality. A Study of Agency and Confederations in Contemporary Outdoor Dance Practices'. PhD thesis: Coventry University, Centre for Dance Research

Noë, Alva (2004): *Action in Perception.* MIT Press

Reeve, Sandra (2008): *The Ecological Body.* PhD thesis: University of Exeter

_____ (2010): 'Reading, Gardening and "Non-Self": *Joged Amerta* and its Emerging Influence on Ecological Somatic Practice'. *Journal of Dance and Somatic Practices* 2(2), 189-203

Paula Kramer is an artist-researcher and movement artist based in Berlin. She holds a practice-as-research PhD in Dance (Coventry University) and was a post-doctoral researcher at Uniarts Helsinki between 2016 and 2019. She currently parents two children whilst being active as an independent artist-researcher as well as, until the end of 2021, a visiting researcher at the Centre for Artistic Research of Uniarts Helsinki.

Her work explores intermateriality through site-specific, outdoor movement practices; rooted in Amerta Movement (Suryodarmo) and Non-stylised and Environmental Movement (Walk of Life/Poynor). She draws on new materialist thought and collaborates with materials of many different orders as active agents in the creation of movement, performance and choreography; as well as daily life practices and sense making. She publishes widely in the context of artistic research through bodily practices and is a board member of the *Journal of Dance and Somatic Practices*.

www.paulakramer.de ~ paula@paulakramer.de

The Choreoauratic Body

Walking as an Act of Recovery

Becca Wood

Abstract

The choreoauratic body is explored in this chapter through tuning into listening, walking as an act of recovery and loosening the borders of body and place through concepts of the nomad and somatic practice. Presented in Dunedin, New Zealand this participatory performance walk, mediated through the headphonic, is described by the author using excerpts from the work and from the participants' feedback, reflections from creating the work and participating in it and through the lens of postmodern and new materialist thinkers. The practice, situated in Aotearoa New Zealand looks to indigenous thinking, acknowledging a Māori world view through resonating ideas of body, site and walking.

Looking west towards the hills. If you look east,
north stretches out along your left arm and south
out of your right finger tips. Take some time on
the bridge to look in both directions, look at
its[the bridge's] structure, the surfaces, its bones, its fluids.
Notice your own soft fleshy surface in relation to
the architecture of the city, and the water of
the Leith. The body and the city merge, the bones
scaffold, the fluids flushing, networking, creating
passages to pass through.
When you feel ready turn towards the south, and follow Forth St
in the direction of the yellow building.
After the bridge there is a carpark on the right.
I can hear buses waiting, lined up, panting, ready
to transport bodies through the arterial routes of
the city.
Head towards the next street corner you will see
a large yellow industrial looking building, its
yellowness stands out against the blue sky.[1]

This chapter explores walking as a performance practice through a participatory choreography that took place in Dunedin, New Zealand in 2013. The work examines the potential for somatically informed choreography to shift our awareness of the environment that we inhabit and to stimulate actions of recovery: recovering forgotten stories of place, recovering from falling/failing and recovering the subjected body in a colonised space through the ability to inhabit the environment through listening. Four ideas are discussed that are all particular to this ambulatory performance and help define the overriding method, which I call choreoauratics:

1.   Listening as a way to 'tune in' to the sensate body
Through reconfiguring the senses towards listening we might tune into the world differently. The act of listening becomes a subversive strategy that rethinks the dominance of the ocular-centric tendencies of Western neo-liberalist culture, towards a heterarchy of the senses.

---

[1] Excerpt from *E-bodies: listen up, tune in, slow down and play on* (see p.61).

2. *Walking as an act of recovery*

As we walk we fail to fall. Each step we take we stumble in multiple, repeated acts of recovery. These choreographed walks became multiple actions of recovery.

3. *The nomadic subject*

The Deleuzian positioning of the sensate body over the productive body is expressed in the nomadic subject. If the terms of a site are also considered nomadic, in accordance with French philosophers Gilles Deleuze and Felix Guattari, there occurs a process of deterritorialisation which loosens the borders of body and also erodes distinctions between body and place.

4. *Somatically informed choreography*

Somatically informed choreography offers cues towards methods that tune the relationship between spaces of the inner body and the surfaces and spaces outside the body, and for becoming more attentive and efficient in activating movement.

Somatic practice (a holistic mind/body practice that spans therapy, training, education, research and creative practice) is conventionally explored either in private studios, or in insular and socially cushioned environments and sometimes in outdoor environments – often these are in natural surroundings. In this work cues adapted from somatic lessons were developed in the score for their potential to facilitate openings in perception through movement praxis as well as processes for attunement to site.

These performance walks shifted through modes of listening and participating and asked: what if we take somatic practice into an urban environment? The use of headphones to help mediate the senses and place became a strategy to enable this work to be experienced in urban landscapes rather than the seclusion and safety of studio spaces and to develop new modes of encounter. The term 'somatically informed choreography' prioritises a greater awareness of the places we inhabit and the way we inhabit them through tuning into the way our bodies might occupy space.

E-bodies: listen up, tune in, slow down and play on

The performance walk *E-bodies: listen up, tune in, slow down and play on,* was framed as a somatic participatory performance encounter, experienced while walking and listening to a sound score through headphones. It was presented at the East West Symposium at The

University of Otago in February 2013. The ambulatory format resisted a typical conference delivery, taking the conference attendees outside into the surrounding urbanscape. The sound on the headphones is composed of a spoken somatic score-based choreography, stories of the site and a merging of environmental field recordings. The term 'choreoauratics' is disentangled through the description of this practice and further defined later in the chapter.

> We set off together as a group on a very warm Dunedin summer day. An organised group of bodies; a collection of flesh and bones, organs and connective tissue, concrete, dirt, brick and water we assembled and disassembled, flocking within the sound score in a kinaesthetic sociality. Sound thresholds merged as the intensities of volume of headphonic sound and live sound competed with one another. The outside space became ambiguous through sound. Was it outside the body, or inside? Or perhaps in the spaces between the bodies? Through these entanglements, borders loosened through sound, time and place, dispersing, connecting and multiplying the 'matter' of this walk.

This sonorous somatic attention generates, as political theorist Jane Bennett calls for in her book *Vibrant Matter*, a political ecology of things, "a cultivated, patient, sensory attentiveness to nonhuman forces operating outside and inside the human body" (2010: viii).

As each participant tunes into listening to the score, there is almost no dialogue between the members of the group. This body collective, all wearing headphones, walk silently. This not only gives the group a kind of performative agency marked by the prosthetic attachments, but also the way they attend to space and one another reinforces a networked space – a silent walking chorus. In sync with the places they walk through, a group sociality emerges out of this mutual connection to site, while simultaneously holding individual headphonic choric spaces.

I watch as bodies find a place to rest in the public site. They soak up the sensations of the day in their skin and in their bones by the way they position their bodies in relation to the urban architecture and the sound score; they seem to be held lightly together by the sound. I am reminded by these bodies, as they shift within a silent chorus inside the city, of Deleuze's interpretation of movement as a translation of space, and through his rethinking of Bergson's expression of duration, as bodies that persist in the landscape "through spatio-temporal forces" (Braidotti, 2006: 3).

By softening the borders between outside and inside through listening and movement in response to the sound score, the experience of the sensate is heightened as is the moving body. Listening, as tested and mediated in this performance walk, seems to have the effect of enlivening the other less dominant senses such as our kinaesthetic senses, smell, touch and proprioception in response to the urban landscape. This allows a different experience of the body in the city, as opposed to the body driven by the productive forces of the social, urban and political codes of the city, which are mostly indicated through dominant visual and spatial references.

Participant's response:
Sound of traffic infiltrates the headphones,
A tin can stutters on the road we are crossing, Wind is tearing the sound from my ears.
The ghosts of songs. Crescendos of wind and voices.
Is the sound recorded or is it the world?

When I first mapped the walk, I took a recording device with me, headphones cupping my ears, microphone balanced gently in my hands. My digital sound recorder simultaneously captured the acoustic world around me as well as my own bodily walking sounds, which were magnified and channelled into my headphones, my ears and then back through my body. In the process of making these field recordings that were to become part of the soundscore, listening to the environmental soundscape through the headphones altered my own perception of place. Some sounds seemed much louder and closer than the ears could hear naturally, making the imperceptible perceptible; the sounds of the textures of clothes moving over the body, the crunch of feet meeting the different surfaces on the ground and the wind beating on the microphone. Digital sound recording processes augmented the environment and the microphone dis-organised the natural acoustic order. The microphone and the headphones recomposed the acoustic selection process, channelling unexpected digitally magnified sounds into the body, through the ear canal, vibrating and unsettling the volume intensities. I noticed details that I might normally miss, my perception of the urban landscape became reconfigured through the headphonic.

Listening in this mediated way opened perception and I became more aware of my breath and my body's walking rhythm. As I perceived my world differently I 'tuned in' to shifting vibrational sonic intensities, my sensation of time and place amplified. I recognised responses that parallel the spiritual meditative processes that feminist writer and philosopher Rosi Braidotti refers to, whereby the "capacity to 'take in' the world, to encounter it, to go toward it" is emphasised (2011: 234). This evokes, as Jane Bennett suggests, "more attentive encounters between people-materialities and thing-materialities" as we experience a difference in attention to our social and situated worlds (2010: x). Listening, somatically informed choreography and the nomadic subject confront our anthropomorphic positioning (a view that prioritises human needs and drives) and as a consequence the binary between human and non-human loses potency.

The drifts in perception experienced in the choreoauratic score became a way of remapping place, through an electro-acoustic field of restless intensities. This was evident not only in my own experience choreographing the walk but also in some of the feedback I received from the participants.

> *Participant's response:*
> *I could not smell anything until I took off my headphones... On the walk back, I could see more than I typically do, but I was aware that busy 'I' had re-entered my mind.*

Her comments seem to indicate that throughout the duration of the walk, the process of listening had relaxed her mind as well as re-composed her perception of the sensate, evident in the change in capacity of her olfactory senses after the walk.

This works towards a reconfiguration of the senses and towards listening as a polemic, which challenges the pre-eminence of looking so that we might tune into the world differently. The focus on vision in the 21st century is emphasised by our daily interactions with technology and screen-based media. The 'commodification of the scopic' is proposed by Braidotti as a driving force in Western culture and a dominant force in the capitalist postmodern economy (2002: 245 & 155). She says that the dominance of the gaze as a driving concern in feminist thinking, "...tends to reinstate a hierarchy of bodily perception which over-privileges vision over other senses, especially touch and sound" (ibid: 246). The acoustic world, Braidotti suggests, is both the most collective and the most pervasive. She refers to artists who engage with the counter-culture of the sonic as a way of "reinforcing different frequencies through technology" or "marking spaces of intensive connection to impersonal and indiscernible others" (Braidotti, 2011: 108).

Listening is positioned as a subversive strategy designed to achieve a heterarchy of the senses. The audio score brings the invisible stories of this place into the choreographic score, entangling the past and present landscapes of the body and the environment. The cocooning of the headphones brings sounds into the body, dissolving the boundaries between the skin and the city. The act of listening is considered a counter practice where listening, as well as the other less dominant senses (such as proprioception and kinaesthetic sensations), are given priority over looking.

"The foot fall sound connects the group", reflects one participant, noticing how they have come together through sound and rhythm. In the moments that the group assemble, their collective presence commands attention from passers-by. There were times when the headphonic participants become unmistakably part of a choreographic ensemble as they converge in movement. As they disperse, their presence becomes less apparent and they dissolve back into the spatial/social configuration of the site. The margins of performing and not performing begin to flex. The participants of the choreoauratic score engage with the urban architecture in a different way from the pedestrians who normally pass through this space; participants pause, slow down, or more radically lie down in the public space. They invoke a playful, anarchic engagement with the urban terrain as they reimagine the city through listening and moving.

> *Participant's response:*
> *I want to walk in the sun but I can't – don't want to break the shared experience, but felt what was perhaps the inevitable pull of the crowd, the group, an urge, a lure to keep walking and following, to be part of that.*

The action of walking together becomes a political act; not only through the Māori view of the term hikoi (an indigenous term for an activist's march), which is discussed later, but in the way the group moves together, clearly marked in their collective appearance, coded by the prosthetic attachment of the headphones, they create a distinct critical mass through how they appear and how they act. The choreoauratic score initiates an engagement with the urban landscape in ways that subvert the pedestrian use of the space. The participants pause in pedestrian pathways, they lie down, they rest, they cluster together. These actions challenge the everyday choreography of the urban architecture, subverting the sociality of the public space, shifting our understanding of a colonial architecture.

Notice how you are standing. Feel the weight
pouring through each leg and down into the
concrete. Stand strong and face the hills.
Feel the distance between you and the hills.
Feel the air on your face. We are going for a walk.
Walk with me. Don't worry, you won't get lost.
Getting lost is how you find your way rest, lie, lean, pause.

Notice your breath, the rhythm of your feet walking, and the way
your bones shift underneath your skin. I follow along with you like a
flock of butterflies brushing your forehead. Lean into the scaffolds of
the city... Fall. Sit. Lie. Perch. Fold. Meander.[2]

Using my body, affected and extended by prosthetic recording and listening devices, the remnants from previous choreoauratic scores are enunciated through the whakapapa[3] of the practice. The remains of my body's movement pathways resurface in these choreoauratic scores, engendering the politics of Rebecca Schneider's *Performing Remains*. Like she suggests, "performance challenges loss" (Schneider, 2011: 102) where the bodily trace remains differently, and through remaining differently there is the potential for change and perhaps recovery rather than remains.

Flax bushes pushed up through the mud, now sealed in by a concrete
layer, and the sea rolled in to lick the foreshore. Reclaimed, a fake skin
surfacing the mud flats. Gigantic birds meandered through the flax in
a time before men. Looking north east across the water of the Leith
which widens and runs out towards the sea, stretch out your arm....[4]

Participant's response:
...but the body does not fall. Each step catches me.
The material of stone, porous, crumbling. The body of the earth,
Holding me, upright.

Working within the notion of walking as recovery, social choreographer Andrew Hewitt presents a choreographic spectrum in which he brings

---

[2] Both paragraphs are excerpts from *E-bodies* (see note 1)

[3] Whakapapa is genealogy, a line of descent from ancestors down to the present day. Whakapapa links people to all other living things, and to the earth and the sky, and it traces the universe back to its origins (Taonui, 2015).

[4] Excerpt from *E-bodies* (see note 1)

everyday movement into the choreographic realm. Stumbling (the failure to fall, or the failure to fail) is, as Hewitt suggests, an "unbelievable power of recovery" (Hewitt, 2005: 89). He identifies the stumble as "not a loss of footing but rather as finding one's feet" (ibid: 89). This brings a politic to the everyday action of walking as a choreography and choreography as an act of recovery. Within these choreoauratic scores we attempt to recover forgotten stories and memories of the pre-colonial and colonial urbanscape we pass through, through the imagination and also through the choreographed walk.

The physical action of walking becomes a threshold *experience* between falling and not falling. Extending this idea within the practice, Braidotti's concept of the threshold is explored through the limits of the subject, whereby the technologised body (the body wearing headphones) goes beyond the limits of the body through mediated sound in the potential multiple tunes of 'others'. According to Deleuze and Guattari, the process of deterritorialisation loosens bodily borders, eroding the distinction between body and place.[5] Reterritorialised by the prosthetic attachment of the headphones, the ear also becomes a threshold, receiving the varying intensities of sound waves that vibrate the interior and exterior spaces. The site and the body become nomadic, a 'smooth space', a distributed space determined by different frequencies, "in the course of one's crossings" (Deleuze and Guattari, 1987: 559).

Braidotti's description of Deleuze and Guattari's text 'Anti-Oedipus' is explored through this practice of choreoauratics in a "joyful anarchy of the senses" in which a reorganising of the sensorial order interrupts monastic and binary thought (Braidotti 2002: 124). This nomadic re-mapping of the mediatised world and the natural world as Braidotti suggests, channels a molecular re-arrangement of intensities and connections in a "cacophony of many insect-like acoustic environments" (ibid: 170). Braidotti's post-human stance offers a synchronous theoretical platform for Elizabeth Grosz's 'species-specific tunes' where there is also emphasis on listening as a counter-position to the ocular-centric drive that dominates the Western regime (Grosz, 2011). If we begin to tune into how other species might hear, perhaps we reimagine the way we hear. In this electronic age, how might we listen from the inside out?

---

[5] "Deterritorialization must be thought of as a perfectly positive power that has degrees of thresholds (epistrata), is always relative, and has reterritorialization as its flipside or complement" (Deleuze and Guattari, 1987: 62).

Participant's response:
I could have stood there much longer... the freedom to lag behind a little, to explore the feel of the leaves, the uneven structure of the walking surface, the hot sun, a bird perching on an electricity post, the mechanical machine, at the hand of its human operator, intent on boring holes into the sealed surfaces of Leith, the birds waiting to fly the dry sweepscape of the uprooted concrete riverbank, a need to place my foot into a concrete foot imprint left behind, to walk around a corner, your voice (in the recording) tapering off with "I lost my leg"...

Headphones cocoon our sound space, prosthetic listening brings the sonic to the inside of the body, reconfiguring the sensate, allowing us to perceive differently. Through these entanglements, borders disintegrate through sound, time and place, dispersing, connecting and multiplying the 'matter' of these walks. This sonorous somatic attention calls for a shift in the way matter is both experienced and considered. Exploring vibrant matter through somatic listening, a body and the site can be said to take on similar qualities. I attempt to bring a different attention to somatic practices where the archaeologies of both the body and the urban landscape become together. Once again, Bennett's attention to "encounters between people-materialities and thing-materialities" reverberates here, bringing a greater sense of responsibility to our anthropomorphic positioning (2010: x).

Choreoauratics brings together the ontologies of somatically informed choreography, the conditions of listening through headphones and critical spatial practice.[6] It's a term that has evolved out of a performance practice that sought to activate the politics of place, language and the body. This is explored in the context of the dispersed, disembodied and accelerated social structures of the highly mediated digital times we are living in. Ambulatory choreographies experienced through headphonic scores look to the conditions of the threshold in urban public spaces, working poetically towards a recovery of the imperceptible and the disappearing. Engaging with the margins of urban spaces, the practice orchestrates an emergent form of public activism.

––––––––––––––

[6] Critical spatial practice is a term that Jane Rendall introduces in order to discuss the operations of art and architecture as they come together in public projects to both question the function of the disciplines themselves and also the social and political relations that these projects create as they engage or critique dominant ideologies (2008: 1).

Acknowledging a Māori world view

Through Māori language, we come to understand land differently. 'Land', in te reo[7] is whenua[8] which also means 'ground', 'placenta' and 'together' (Mead, 2013: 15). Choreoauratics resonates with a Māori world view, where the borders of land and body are reconsidered from conventional colonial perspectives. Later in the same year as the work was made, and while in Dunedin at the Aotearoa Digital Artists Symposium, local Māori orator Tuari Potiki spoke of the importance of 'making visible' the places of the tangata whenua[9]. His stance resonates with my own concerns in this practice, rethinking our experience of the contemporary city through existing networks and memories (Aotearoa Digital Artists Network 2013). Potiki recalled the histories that have disappeared, the pre-colonial stories that might be recovered as we 'take place' in reimagining the city for the future. A persistent undercurrent in my practice was to make visible that which has become invisible and to recover the lost or disappearing.

Hikoi is the Māori word for walk. Hikoi activates walking *and* talking and suggests a commitment to movement, to community (coming together) and to connection to the land (Barnes, 2009: 7). Hikoi, like Hewitt's provocation, also presents the mechanics of a society, our movement through space and on land and how we do this together. In contemporary times hikoi has come to be understood as a way of taking action against or for a specific political or cultural view.

Through performance walks such as these we might practise the cultivation of a slow attentiveness to non-human forces that operate both in and outside the human body, where the unspeakable is expressed through forces of movement, where we might reconsider notions of loss and stumble to find our feet in a recovery of what remains.

---

[7] Te Reo Māori is the indigenous language of Aotearoa, New Zealand.

[8] Whenua is the Māori word for land. It also means placenta, "All life is seen as being born from the womb of Papatūānuku, under the sea" (Royal, 2007).

[9] Tangata translates as 'human beings' or 'persons'. Tangata whenua means people of the land.

References

Barnes, H M. (2009). *The Evaluation Hikoi: A Maori Overview of Programme Evaluation.* Shore and Whariki Research Centre, Massey University

Bennett, J. (2010) *Vibrant Matter: A Political Ecology of Things.* Duke Univ. Press

Braidotti, R. (2002) *Metamorphoses.* Blackwell Publishers

_____ (2011) *Nomadic Theory: The Portable Rosi Braidotti.* Columbia University Press

Curtis, M. (2016) 'The Poetics of Bilanguaging: An Unfurling Literacy; Ngā Toikupu o Ngā Reo Taharua: e Tākiri ana te Aroā Pānui, Kamatekaora', *A New Zealand Journal of Poetry and Poetics 14.* Online at https://bit.ly/banda25

Deleuze, G. & Guattari, F. (1987) *A Thousand Plateaus.* Univ. of Minnesota Press

Grosz, E. (1995) *Space, Time and Perversion.* Routledge

_____ (2011) *Becoming Undone.* Duke University Press

Hewitt, A. (2005). *Social Choreography: Ideology as Performance in Dance and Everyday Movement.* Duke University Press

Mead, H. M. (2013) *Tikanga Māori: Living By Māori Values.* Huia Publishers

Potiki, T. (2013) *Where We Stand* in *Space: Network: Memory* – Aoteroa Digital Artists Symposium 13 September 2013, Dunedin, New Zealand

Rendall, J. (2008) *Critical Spatial Practice.* Denmark: Art Incorporated, Kunstmuseet Koge Skitsesamling

Royal, T. C. (2007) 'Papatūānuku – the land – Whenua – the placenta', Te Ara – the Encyclopedia of New Zealand, online at https://bit.ly/banda26

Schneider, R. (2011) *Performing Remains.* Routledge

Taonui, R. 'Whakapapa – genealogy', Te Ara – the Encyclopedia of New Zealand, online at https://bit.ly/banda27

Becca Wood has been working in performance practices that slip between bodily, spatial and digital environments for the past 25 years. She coined the term 'choreoauratics', a practice that fuses somatically informed choreography and sonic investigations with philosophies of listening, the body, place, digital technologies and sociality. Framed as critical spatial practice, theories and codes of space and place, the body, and digital technologies intersect to imagine new possibilities in inter-modal performance arts. At the time of writing Becca is lecturing at the School of Performing and Screen Arts, Creative Industries, Unitec following a brief time living and working in the UK in Dance at Coventry University.

becca@beccawood.co.nz

The Dreamweaving Body

Sarah Hyde

Abstract

In this chapter I describe one of the frameworks that I use to increase both physical and environmental awareness. The process includes breathing and movement practices to wake up the body, together with a ritual journey through a particular landscape that facilitates reconnection with the dream of nature.

The 'Wood Element Day' workshop, held in spring and outlined below, is part of an annual cycle called 'Moving with the Elements'. Based on the Five Elements in Chinese Medicine, each day is designed to help participants realign to the natural rhythms of the season and to experience the changes that connecting with that element brings. We follow the process of the group and, in the afternoon, one individual's dream journey.

Introduction

My approach to body awareness and healing stems from many years of practising as a Naturopath, Osteopath, Plant Spirit Medicine and Amerta Movement practitioner. Whilst new clients may come to the clinic expecting a more traditional naturopathic and couch-based approach, I introduce them to basic breathing and relaxation exercises as well as postural awareness in the early stages of their treatment, alongside any bodywork they may receive. This enables them to continue the programme of treatment at home and further develop their sense of body awareness.

When the treatment has progressed to a point where both I and the client feel that they are ready to move off the couch and to take a more proactive role in their healing, we change the emphasis of the sessions from hands-on bodywork to more breathing and movement exercises. This, together with developing a basic understanding of anatomy, increases confidence in how the body works and what it is trying to convey. It is an important step towards changing the habitual patterns of movement and behaviour that potentially are throwing the client out of balance and alignment.

The early stages of this movement practice usually take place in the treatment room or studio, however I have found that it is equally important to continue this practice outside in the natural environment. Being in nature realigns our bodies to the more natural rhythms that modern life is constantly drawing us away from. In this chapter I focus on a journey through the Wood Element Day. These days are based on the Chinese Five Element system, in which each element governs a season and relates to particular qualities in both the external environment and within the inner landscape of body, mind and spirit.

Wood Element Day

Morning Session

> The italicised text in the description of the morning session is my own response to the work, primarily my body speaking from within the dreaming of the exercises. It weaves together the store of body memory I have gathered over the years, my observations of these patterns happening in participants and the feedback they have given. In this way, I can speak from within the body rather than from observing it.

It is a beautiful, clear, Spring morning, cloudless blue sky, warm, birds singing, bluebells in blossom and daffodils coming to the end of this year's blooming. The day is happening on the land where I live, which provides many different environments for movement exploration. We have gathered on the movement lawn, a wide open space, bordered by trees and a broad river on one side with trees and a road on the other. Beyond the lawn is a wildflower meadow, above that a range of hills reaches to the horizon. For a while we sit in silence, feasting our senses on the nature surrounding us. The beauty of the morning is shattered at intervals by the roaring of motorbikes heading for the mountain pass, and by the sounds of a chainsaw across the river.

I begin the session with a short introduction describing the qualities of the Wood Element and the season it relates to – Spring. Some of these qualities are:

- The creative force
- The vision and plan needed for a seed to achieve its growth to fruition
- Growth
- Boundaries
- The emotion of anger[1]

I then set up the following exercises designed to begin waking up the body, specifically the systems within the body that are governed by the Wood Element: the liver and gall bladder.

Walking Exercise

This is simply walking barefoot on the grass. At intervals I suggest some different sensations and experiences on which to focus awareness.

- Feel the feet meeting the soft cool grass and sinking down to meet the resistance of the earth.
- Does foot sink into the earth or does it feel like the earth rises up to meet the foot?
- Bring attention to the meeting place of sole and earth where both meld together.

---

[1] For a more in-depth description of the elements, please refer to my bibliography.

- Feel the movement between the bones of the feet and in the flesh and tissues that surround them.
- How is the weight falling through the feet?
- Where is it falling and what imprint would this leave on the ground?
- How can we stay interested in the meeting of feet with the land as our eyes see a red kite fly overhead, our ears hear the river singing on the stones, or our whole body is jarred by the roar of a motorbike?

Lying, Rolling, Crawling, Standing, Walking

I follow the walking exercise with the above sequence. For those not used to lying on the ground outside this is an interesting process in itself and I often witness initial resistance to this, both in myself when I practise alone or with students.

> I have begun to slow down a little with the walking exercise but my chest is still tight, breathing restricted, mind racing from countless things to do. My vision and brain are fuzzy. I feel very irritated by the intermittent sounds of chainsaw and motorbikes. They sound so aggressive and destructive, especially as I am trying to become more sensitive to my body. I long to lie down on the ground but my mind says it's damp with dew, muddy bits, I'll get wet. I squat, put my hands in the grass, its fresh aroma fills my nostrils. I drop beneath the voice that says 'don't' and lie down on the ground. An absolute sense of relief as belly meets earth. The wet grass is refreshing on my face, the sun comes out from behind the trees and warms my back, I hear the river singing, my breathing slows, chest and belly soften and a whoosh of deep full breath opens up my being to a new depth, then releases back to the earth. I could stay here forever, just dropping deeper and deeper into myself, into the earth. Then crash, through the deep peace, the roar of another motorbike. Interestingly, despite its vibration ricocheting through my body, I am able to re-engage more quickly in the exercise, bringing my attention back to my own breathing. The feeling of aggression that inhabited my body dissipating quite quickly...

It is very easy to want to remain in the lying stage. How to give the time needed for this and then gently introduce the next stage of movement is an important part of opening to greater body awareness. When participants arrive in this place in themselves, if they jump out to do the next exercise too quickly they can revert to their habitual rhythm rather than come from the more natural rhythm they are beginning to access. My guidelines to aid this process are:

- Gently move with the breath
- Exaggerate the movement of the breath
- Roll slowly from side to side
- Feel the sides of the body on the earth, in the air, the comparison of the solidity and containment on one side, and the space and openness on the other.

...as I lie on my side feeling my ribcage expanding and relaxing I realise I am not in the habit of being aware of my sides, the rhythmic pumping of my heart, the expanding and relaxing of my lungs, all these movements that are continuous within me night and day. I see a spider crawling up a blade of grass, down the other side, then up another stem, I find myself turning over to watch it more easily, enjoy the feeling of rolling and turn again. Suddenly I am rolling over and over on the lawn and a feeling of freedom and childlike delight fills my being as I do so.

Becoming more aware of our bodies we notice how our senses are continually picking up messages. How we receive these messages can help or hinder the waking up process. Becoming lost in a sound, sight, or smell without remembering our own body, leaves the latter 'empty' as our attention has gone out to the stimulus. This leaves us very vulnerable if a different sensation appears, particularly an unpleasant one. The example above shows how being more present in the body enables us to rebalance ourselves more easily when disturbed.

The emotion of the Wood Element is anger. It is healthy to observe that when we experience a boundary violation, anger naturally arises as a way of re-affirming the boundary. In the example above the irritation arising in response to the motorbike is channelled into a greater focus of attention on the breathing and movement practice and this change of emphasis restores the physical boundary. In the plant world, when a seed sends up shoots, if these meet a rock they try to find a new direction around the rock; this is a healthy Wood response to a boundary, being adaptable and able to change direction when necessary.

Lying on my back, feeling the warmth of the sun on my body, I hear a crow caw as it chases a screeching kite. My eyes immediately search the sky for the birds. I hear the instruction to receive the sound and image of the bird rather than shoot out of my body along the pathway of my vision. As I breathe and feel my back on the ground I have a very different sense of seeing, it feels less grasping, my head feels softer, my eyes gentler. I remember myself and am less inert.

Later in the morning the chain sawing starts up again. It is a continuous discordant sound, unpleasant and jarring to the system. As facilitator I wonder how I can continue the day with this noisy onslaught. I am also aware that some of the group are getting very distracted and irritated by the noise. Eventually, however, this proves to be a great teaching tool for introducing the concept of boundary issues as a body awareness exercise. If we feel a sound is invading our body, we need to move strongly in order to push the sound back; this way our own energy expands rather than contracts and the sound can be held at our boundary. Or we can relax internally, allow the sound to pass through our body and release it, without creating discord.

The following are some of the movement suggestions I offered to help individuals rebalance their boundaries in response to the noise:

- Move strongly with your legs. (To me the legs represent our sense of direction and purpose and movement practice that focuses on them can help to free up the liver and gall bladder meridians by moving the qi along them in a positive way.)
- Stretch or focus on the movement in the sides of your body. (This gives more flexibility to the ribcage and opens up the chest and abdomen, massages and gives space to the liver and gall bladder.)
- Use the voice, particularly give a good shout, as this is a wonderful way to expand the energy through the body (the sound of the Wood Element is 'shout').

I observe a complete change of energy as the group focus on these movement tasks. The sound of the chainsaw continues, but it recedes into the background for increasingly longer intervals. Despite the disturbance, each individual is developing more ability to keep to his/her own line.

Near/Far exercise

The sense organ of Wood is the eyes – our vision. Having clear vision, physiologically, mentally and spiritually is a good indicator that the liver is functioning well. The following exercise is to help bring more awareness to our seeing.

I suggest the group allow their eyes to rest on a distant view such as the sky and observe what happens in their bodies as they focus on this.

> ....as I became absorbed in the canvas of the sky, the wisps of cloud passing overhead, a vapour trail carving a straight line across the blue, I begin to feel a trace of a headache. I bring my attention back to my body and my eyes feel bulgy with the tension of the little muscles that surround them, my jaw has tightened, as has the back of my neck, I am breathing more shallowly. I focus on relaxing my face, breathing out through my mouth to try and soften my jaw and neck, take long deep breaths in and out to open up my chest and allow a sense of inner volume to return. I focus on my hearing a little in order to rest my eyes. From this more relaxed place I look at the sky again. This time I stay on the ground and allow the images to come to me rather than me 'going out' to look at them, it is much more restful like this and spacious. The headache has receded....

I then suggest they focus on their near surroundings

> ...as I look at the near surroundings I become absorbed in the detail of a dewdrop, the antics of a beetle and an ant on the same blade of grass, my fingers and toes, elbow and knee.... I stop, hold my position, feel my shape, breathe, see my surroundings then continue. I become part of this landscape rather than looking at it....

After a break I demonstrate how we can make movement sequences out of the morning's practice. The participants begin to move between different levels of lying, crawling, standing and connect their movement with their seeing of near and far. I observe how, as they put these exercises together, their movement vocabulary expands. At the beginning of the morning, sequences of movement had been short before each person would stop, unsure what to do next. Now, with more vocabulary they are able to develop much longer, more fluid movement sequences.

{}

Afternoon session

The italicised texts in the afternoon session are extracts from one participant's journey that was written up from their notes, several days later.

After waking up the body in the morning, the afternoon focuses on a journey through both the physical and imaginary landscape.

There is a place on our land that is perfect for this. It is a little gorge with a stream that flows down a cascading waterfall, which then flows down into the main river, joining at right angles. The banks of the gorge are wooded, the trees creating a green canopy overhead, their moss-covered roots intertwining with the rocks on the riverbed. Access to the gorge and waterfall is through an orchard, the path leading down through a gap between upright trunks of oak trees. On the first ledge down, embedded in the earth, is a large stone with numerous glistening quartz crystals.

My instructions for the journey are:

- Take time in the orchard to walk around, settle after lunch, relax your mind and allow your intention for the journey to surface. This may be in the form of a question you need help with.
- Once you have your intent, come to the doorway to the gorge.
- Stop for a moment and be mindful as you step through the doorway.
- You are now a guest in this Elemental world. Please ask permission to be there and wait for the answer in whatever form this comes.
- Repeat your intent to yourself, before letting it go.
- Whatever happens in your journey by the waterfall and on down to the main river, trust that it is relevant to your intent.
- Allow yourself to go where you are called or feel; to sit where you need; give yourself time to enjoy being in this space, using the bodywork of the morning to help you stay present.
- If you come to a place where you would like to spend time and other beings such as plants are already there, give them priority. Remember, you are the guest here in this time and place.
- When you reach the confluence of the stream and the main river, give thanks to the gorge and the beings that live there for what they have mirrored back to you.
- Spend time at the confluence before coming back up to the orchard. On your return, find a quiet place to write notes on your journey.

M's journey

The journey was two weeks after telling my partner T. that I wanted our relationship to end.

My question/intent – 'how do I be around T. at the moment?'

First, I noticed the contrast between the bright orchard and the dark slope into the unknown. I felt a sense of anxiety, foreboding, real fear of the treacherous, smelly and damp. I was just wandering to start with... a small fight going on between my body and my mind. Body not wanting to be in this dark place, mind not wanting to be in my dark thoughts. I noticed how tight my jaw was.

It is often the case that our perception of the external environment mirrors how we feel inside. If we can understand this, it empowers us to make adjustments within or without, depending on what we feel we need to change.

...So I looked for small places that I could make a connection with. I found a very round pool, but that was murky and stagnant. Then a deep cut in the rock that was sunlit, but shockingly cold. The stones felt slippery... I had a strong sense of just wanting to feel safe, and that grew like a primal and urgent need.

Here M. has looked externally for a place of safety. Then a little plant spirit catches her attention. The effect is immediate, its message so clearly mirroring back to M. her inner needs at this time:

Then I noticed a tiny silver fungus growing from a tree, like a ball. I touched it gently and was amazed by just how silky delicate it was. So tender, it could only take the softest of touches. I felt the years of T's rough treatment of me and an ache for how much I wanted this kind of respect and gentleness. I so wanted to stroke it, but then realised that even my gentlest touch was starting to lift its skin. Now I felt a stab of horror and guilt. I apologised to it for taking too much, I wished it healing and I left it to live its life without my intervention.

It is easy to take for granted the compassion of the Elemental world and how it tries to communicate with us. Yet when we stop and listen, as M. does to the silver fungus, there is a potential for dialogue. This may be in a language beyond the usual 'understanding' of our minds but it is of no lesser importance. This interchange brings us nearer to a place we used to naturally inhabit and be a part of, and gives a great sense of wellbeing.

When we meet this world with more awareness, in our body, feeling and senses, we bring life into this relationship, rather than just absorbing from it. In a similar way, we bring life to our own bodies by becoming more aware of ourselves.

> ...I noticed how clumsy-footed I felt in the water and on the stones, I sought a stick to support me. I found one and held it perpendicular in the water, noticing how the current would arc round it. For some considerable time I just enjoyed sensing the water's determined energy, almost like it was gently correcting me again and again. Like it was saying 'no, this is the way it should be'.

> As I got more attuned to the water and its energy, I started sending small sticks down in the current. I followed each one's journey, watched how they could get stuck and then come loose again, joyfully bobbing and careering their way downstream. I felt a real visceral joy in their freedom.

As M. becomes more engaged with the Elemental world, she finds herself taking part in a story that later she realises is mirroring back the answer to her intent. For now, though, it is important to just trust the unfolding of this story and allow it to reveal its message. No amount of trying to make something happen in this realm works.

> My attention then got drawn to two sticks that had got stuck. In their trembling, they seemed to be aching to get loose. I was really empathising with them! So I started sending other sticks down the river to see if they'd get free (I call these sticks the 'dislodgers'). One of the two sticks got dislodged easily and I revelled in seeing it bouncing and skipping down into the sunlight. I felt a sense of myself and my ending of a difficult relationship. Through all the guilt and worry, I suddenly felt the joy and hope of my own freedom.

> My attention then stayed with the other stick, still trembling for release. Many of my 'dislodgers' simply bypassed it, and one even went under it and out, without impact. No stick would take it. I felt (its) pain on its behalf, but eventually I decided I had to move on and it's a bittersweet feeling to leave it there and head out down into the sun. I felt my own joy in joining the other sticks that had been set free, but also grief for the stick I'd left behind. I sat in the sun and revelled at these two feelings, delight and hope, as well as grief for my ex-partner's defended stuckness. My body now felt soft and kind, no jaw tightness.

When the participants have assembled back on the lawn, I suggest they each find movements that depict the beginning, middle and end of their journey, and to put these together in a movement sequence. This becomes a ritual as each participant performs their sequence to the group before giving any verbal feedback. After feedback they close with a repetition of the movement sequence, consolidating the experience of the journey in the body.

> ...Back with the group, when I rework this journey in movement, I play out the moment when 'my stick' is freed. I do it too fast and strain a muscle. I reconsider the moment of freedom and actually, in the river, it was not a violent release, it was sure but soft. The actual moment of freedom is a tiny one. A stick released by another stick's touch is gentle, just a tiny nudge, a glancing contact. The first moment of freedom is so subtle, just a tiny turning. And then the stick slowly arcs around to join the energy of the river...

When our habits take over the body's own telling of the story they can cause a discord that previously was absent.

> ...The journey gives me a new somatic sense of the delight of slowness. I recognise how in life my speed can hurt me. In further explorations after the workshop I acknowledge how my life history taught me to speed up my actions, decisions, processes, out of panic. Actually (just like the little silver fungus) this sensitive slowness is what I need. In my disentangling from my relationship with my ex, I've continued to embody this somatic awareness which I gained from this journey.

The journey from the couch-based treatment along the spectrum of learning to listen to the body through breathwork and gentle movement exercises, on to the outdoors, further developing one's movement vocabulary and reconnection with nature, is one of self-healing and so one of self-empowerment too.

The body's innate ability to heal itself, given the right parameters, never ceases to amaze me. Symptoms that were so persistent in the treatment room, seem to release their hold as the mover surrenders their customary tension patterns and softens into a stiller, gentler, more compassionate place within themselves that is far more in balance with the natural rhythms of life. And, most importantly from my perspective as a practitioner, self-generated.

I do believe that the more aware we become of our bodies, our thoughts and feelings, the more chance we have of understanding what they are trying to say to us. With this understanding we develop more awareness of what throws us out of balance and how to make choices in our daily life that enable us to redress those imbalances. The sense of wellbeing that is cultivated benefits not only ourselves but extends to the environment we live in because we lessen the discord we create within it.

Acknowledgements

I would like to acknowledge the feedback I have received from Fiona Latus and Nick Sales in developing this chapter. And to offer my gratitude for the teachings of John Garrie Rōshi and Suprapto Suryodarmo.

Further Reading

Bloom, K., Galanter, M. and Reeve, S. (eds.) (2014) *Embodied Lives: Reflections on the Influence of Suprapto Suryodarmo and Amerta Movement.* Triarchy Press

Cowan, E. (2014) *Plant Spirit Medicine: A Journey into the Healing Wisdom of Plants.* Sounds True

Garrie, J. (1998) *The Way is Without Flaw: Teachings of John Garrie Rōshi*

Reeve, S. (2011) *Nine Ways of Seeing a Body.* Triarchy Press

_____ (2013) *Body and Performance.* Triarchy Press

Stirk, J. (2015) *The Original Body – Primal Movement for Yoga Teachers.* Handspring Publishing

---

Sarah Hyde received a BSc in Ecology and Zoology from Exeter University before going on to train at the British School of Naturopathy and Osteopathy. Alongside her clinic practice she promotes self-healing through teaching movement and meditation. Her 'Healing Moves' programme in West Wales focuses on how to bring healing and vitality back to the ever-depleting natural environment, both within the landscape and the human body. Her practice is grounded in the re-wilding of the smallholding where she lives with her husband and daughter, dogs and cat.

sarahlhyde@yahoo.co.uk ~ www.sarah-hyde.co.uk

The Emanation Body

How the Buddhist concept of the three kayas helps me find congruence as a therapist, through mind, feeling and body

Anna Murray-Preece

Abstract

A description of my experience of using the Buddhist concept of the three kayas to inform my work as a therapist. The kayas give us a sophisticated way of understanding our awareness of our mind, emotions, feelings and the body that I find helpful in tracking both my own process and that of the client. I focus in particular on how I have worked with clients for whom being in relation to their body is especially difficult, referencing recent understanding of dissociation, trauma and the limbic system of the brain. Through this I demonstrate how it is possible through awareness and movement to support them to land more fully in their body, in a way that is safe and gentle.

I am high on the moor, river below and the sea in the distance. The wind gusts around me, cool and refreshing, a strong raging force, I can hardly stand up.

I stretch out my arms, resting on air, I feel the joy of a wind that can hold my whole body, a delicious sensation of being held and supported.

It flutters and changes, now suddenly it's calm, all momentum is gone. My body adjusts with a sharp downward movement, I must stay awake as the next gust could topple me.

As I root into the ground, gradually I open, attuning to the changes with a flexible body.

Responding to the dialogue with each shift of air, gradually becoming part of the whole.

I move, I am moved, exploring the space, I receive what's around me.

For many years I have had this practice of moving with the wind. It has been an invaluable teacher in my work, encompassing psychotherapy, Plant Spirit Medicine and healing through body awareness and movement. It shows me, increasingly, how I can stay present, while still grounded and congruent with my own energy, supporting me to meet unknown forces: subtle movements of air on one day can be wild and strong on another. It has also helped me in my own journey of embodiment, giving me a way to practise staying rooted to the ground while in relationship to what is around me. In this way it has deepened my experience of embodiment. As a therapist I am more able to attune to the qualities of my clients, whatever they bring, from the most delicate feelings to a strong, angry rage. I can allow my own sensitivity, while being less likely to be overwhelmed by their process.

The Three Kayas

Moving with the wind also reflects three things that are central to my work. The cultivation of spacious awareness, the movement of energy and a relationship to the body. These aspects are echoed and informed by my understanding of what are known as the three kayas in Tibetan Vajrayana Buddhism: dharmakaya, sambhogakaya and nirmanakaya. **Dharmakaya:** (dar-ma-kaya) or the 'truth body', is the quality of mind, or wisdom that realises the empty, spacious nature of reality; **Sambhogakaya**: (sam-bog-a-kaya) or the 'energy body', relates to our feeling, emotional and imaginal life, and **Nirmanakaya**: (near-man-a-kaya) the 'emanation body,' is the vessel which holds these two and as such is the vehicle for manifestation, the physical body.

In the Tibetan Tradition it is understood that our relationship to the kayas is a process that becomes increasingly subtle, eventually becoming the three bodies of a buddha's awakened nature. The kayas, therefore, reflect an expression of the emergence from emptiness to subtle energy and into form. Ultimately, they bring an awareness of the underlying spaciousness of our reality which enables the flow between these three aspects. Out of the space of emptiness comes the wind, which is felt and responded to, then grounded through the movement and expression of the body.

To me the kayas provide a subtle and perceptive model of embodiment, which I have found to be supportive in tracking the process of health and healing in my interactions with my clients. The kayas are reflected all the time in our lives, from the spark of an idea that comes 'out of the blue' through to the process of its gestation and manifestation into form. It is when this movement of energy is activated, but becomes stuck, that we see disturbance in the form of emotional or mental distress and physical disease.

In my practice, I work with a range of people, some who comfortably inhabit their body and others for whom the trauma they have experienced makes this very difficult. In this chapter I will show how the kayas inform me in how I meet each person, allowing them to be where they are, while supporting them, gradually, to land in their body. I will use the example of my client Ellen and her journey into increasing capacity for embodiment.

Dharmakaya – Spacious Awareness

Dharmakaya in its gross aspect is our conceptual mind. When this is disturbed we see the cycling of unwanted thoughts, anxieties or fixed views. As it becomes more refined it is a quality of spacious awareness that can be simply present with what is.

In my work my intention is to allow my mind to soften and relax into spacious awareness as much as I am able. This is where my willingness to let go into the holding of a wider field of awareness, or the transpersonal, becomes an important support. A simple prayer or intention makes all the difference to how much I can let go and open. This helps me to 'let things be', so that my ability to be with another's experience increases. Sometimes there is nothing to be done but simply be present.

> Ellen was extremely distressed, the whole of her known world having fallen apart. How could I be with her with the feelings of extreme grief, fear and loss that were arising? The intensity of her feelings touched me deeply, but I feared I would not know what to say or do. I was, however, able to stay with a sense of spaciousness inside and simply remain present. I felt like a lightning conductor, holding as steady and grounded a place as possible, enabling her to feel heard and received in her suffering. By the end of the session, although nothing had changed in her external world, Ellen expressed that being met in her experience made it more tolerable.

If, however, I contract or solidify a feeling or thought, I may try to fix what is perceived to be a 'problem', rather than allowing it to be as it is. If the mind contracts and becomes caught in a story, the story can begin to define how we feel. This may provide safety to our sense of identity, but it can also stop us from being with the discomfort of our feelings. When we can allow the mind to relax and open to whatever is arising it helps us settle into the underlying energy of sambhogakaya.

The 3 Kayas

Movement from subtle to gross

	Dharmakaya	Sambhogakaya	Nirmanakaya
Human condition	Conceptual Confusion Restless mind	Feeling Emotion Imagination Fantasy	Physical body Structure Sickness Discomfort
Awakened nature	Clarity Spaciousness Consciousness (Truth Body)	Subtle body Bliss, movement of energy (Energy Body)	Vehicle of manifestation (Emanation Body)

Diagram from *The Psychology of Buddhist Tantra* (Preece, 2006)

Sambhogakaya – energy and emotion – the bridge

Sambhogakaya in its gross state is emotional process. When this becomes stuck or over-activated it can pull us this way and that, like the wind on the moor. As our awareness becomes more subtle, we contact the feeling beneath the emotion and the felt sense.

Sambhogakaya relates to what are known as the *energy-winds*, which underlie our feelings and emotions. When they are quiet and still I notice them simply as the energetic movements or feelings that happen in response to situations, relationships and events in my life. When triggered they become stirred up, creating emotion. In themselves our emotions are not a problem, they are a healthy response to what we meet in our lives. It is when they overwhelm us or become blocked that they are problematic.

As the 'energy-body', Sambhogakaya is also the territory of dreams and visions. It is a subtle vitality that is not fixed, but fluid and changeable, where transformation can happen. In terms of density, as energy, sambhogakaya sits between the empty spaciousness of dharmakaya and the form of nirmanakaya. In this way it is a bridge between them. It allows what is unconscious and often held in the body to come into consciousness through our dreams, visions, fantasies and feelings. "… most of the work happens in this middle territory, the subtler territory, where we are less separate and yet fully present" (Sills, 2000). Being in touch with this less separate level of awareness increases my capacity for empathy.

> *Ellen has received shocking news. She is in a frozen state, dissociated from her feelings and her body. As I attune to my body, I feel an ache in my heart, which eventually I decide to voice. She says this is how it is for her also and asks that I continue expressing my feelings as it helps her to connect. I describe the feeling of darkness washing through me. It is almost unbearable, but awareness of my body and contact with the ground helps me to stay with it. As we continue she becomes increasingly open to her feelings and eventually both of us are in tears.*

This was a profound opening for Ellen and is a good example of how an awareness of my physical body enables me to open more fully to the client's experience, in this case to her feeling and emotional life, or Sambhogakaya. It also shows how my capacity to tolerate these feelings helped Ellen digest what was previously an overwhelming emotion. The stability of my body, or nirmanakaya, helped me stay with this process.

Nirmanakaya – vehicle of our embodied expression – 'the emanation body'

Nirmanakaya in its gross state is the pain and suffering that our body can experience. It is the aspect of our being that both contains or holds the other two, mind and feelings, but also expresses them. It is the vehicle through which we can manifest in our lives.

Our body reflects who we are and who we have become in response to the many experiences in our life. This is seen clearly in the work of Ron Kurtz, creator of the Hakomi method of therapy, where he describes character styles that map different ways of being in the body (Kurtz, 1990). One example of this would be the 'sensitive/withdrawn' character who in response to feelings of early childhood rejection can struggle to inhabit their body. We all develop patterns of holding or tension in response to what we meet in our lives. Working with Ellen to embody and move with her emotional process helped her to release some of this tension.

> *Ellen clenches her fists, pressing her thumbs hard against her hands which, she says, helps her manage the sensations she feels in her body. 'I feel wrong'. I suggest she follow the movement of her hands, allowing them to guide her. Tentatively, she moves her wrists, exploring how it feels to bring more expression to the movement. To help her stay connected to her body I encourage her to feel her spine and her feet, I feel a sense of relief in my body as she drops more into her feelings. Eventually I suggest she looks at the hand movements, to receive their clarity and grace, to allow them to play and let the artist in her express herself. Working with her hands seems a good way for her to connect with her body, it doesn't take her too deeply into her trauma but brings her into greater relationship with herself. She is discovering how to be with her feelings without them overwhelming her. As she continues, her movement becomes increasingly integrated, including not just her hands but her whole body. She says she can now feel her heart and her feelings, she is much more embodied.*

Ellen's movement opened a channel, enabling a flow between the kayas from her body to her feeling and into expression. She is now able to witness, rather than be unconsciously driven by the feeling of 'wrongness'. She is increasingly finding safety in her body so that it is becoming a resource that holds her awareness of feeling and cognition. Through looking at her movement and receiving its quality she is beginning to change her perception of herself.

Tibetan Buddhist scholar and meditation master, Dzogchen Ponlop, in his book *Wild Awakening,* says of nirmanakaya that at its most refined "unobstructed awareness manifests as body". (Ponlop, 2003: 140) This points to a condition of being where contracted thoughts, feelings and body processes can release and we are free to express and manifest ourselves with unobstructed fluidity.

I find it remarkable that attuning to the sensations in my physical body leads me to the felt sense and the energetic movements of sambhogakaya. When I can rest in this, my awareness becomes increasingly subtle and still, ultimately taking me towards dharmakaya and emptiness. Meditation teacher Reggie Ray says "The more you explore your physical body, the more it dissolves into energy, and you realise that even the idea of having a physical body is mistaken. The body is an energetic phenomenon onto which we have superimposed the idea of solidity" (Ray *et al.,* 2009).

The body, as nirmanakaya, therefore, is a reflection of where the qualities of our mind and emotions flow or are constrained. For many, especially where there has been trauma, being in connection with the flow of the body and feelings is not always easy. This can lead towards contracted states of mind and feeling, often resulting in dissociation from the body.

Dissociation and Trauma

The mechanism of dissociation, or removing awareness from the body, is an essential protective response to a situation where there may be intolerable pain or trauma on a mental, emotional or physical level. Ideally, when circumstances are safe enough, awareness begins to return to the body. In his book *Waking the Tiger,* trauma therapist Peter Levine describes how the impala freezes in the face of a deathly attack, tricking her attacker into believing she is already dead. When the threat is gone, she gets up and shakes, returning to her body. The shaking releases tension, completing the cycle of the previously mobilised fight, flight and freeze mechanism, leaving her free from any residue of trauma (Levine & Frederick, 1997: 15). Humans also have this facility to move through and recover from traumatic events. If, however, the situation still feels unsafe, the process will not unfold and the energy that has been aroused becomes stuck in fight, flight or freeze. Until this can be released, the body continues to behave as if it is threatened, especially when triggered by an event similar to that which initially stimulated the dissociation. For many people, this means either a tendency to dissociate, or a continual sense of being disconnected from the body, as we have seen with Ellen.

From the perspective of the kayas we could say that unless there is a flow, from awareness or the cognitive level (dharmakaya), to the emotional, energetic or feeling response (sambhogakaya), to being felt or expressed in the body (nirmanakaya), then there will be a block in the process. For healing to happen this flow needs to be re-established. Reconnecting with the body is an essential part of the healing process. In his book *The Body Keeps The Score*, trauma therapist Bessel van der Kolk says "Trauma victims cannot recover until they become familiar with and befriend the sensations in their bodies" (2014: 100). When this happens, the body can then 'hold' the thoughts, feelings and emotions so that they can be digested, expressed or released.

As a therapist I feel my role is to facilitate this flow between the kayas, which can support the release of what is stuck or held.

We cannot, however, instantly jump from a dissociated state in relation to the body, to re-inhabiting it. When we do there is a danger of re-traumatising and entrenching the trauma more deeply. This process needs to be done gradually and carefully. My intention, therefore, is to meet each person where they are, be it in their mind, feelings or body to support a gradual flow between all three so that they become increasingly congruent with each other. I do this with as much awareness as I am able to bring to the process. Bringing attention to the kayas opens me to the potential of the mind and its spaciousness, the emotions and their capacity for transformation and the body in whatever expression it is able or needs to make, in response to what is happening.

If I can sustain this awareness, it can bring space and safety into the relationship with my client, enabling them to have more tolerance of what is happening for them. The energetic 'field' of the therapist has a profound effect on the therapeutic relationship, as reflected in recent studies in neuroscience on limbic resonance.

Limbic Resonance

Limbic resonance is the process whereby our resonance with others can impact, adjust or balance our physiology. Numerous studies show that for our health and survival we need others to help us regulate. In their book *A General Theory of Love*, Lewis, Amini and Lannon (2000) say that "The reciprocal process happens simultaneously: the first person regulates the physiology of the second, even as he himself is regulated. Neither is a functioning whole on his own; each has open loops that only someone else can complete."

This shows us that what we 'emanate' is crucial in defining how we affect others. If I can rest in the stability of my body with a quality of compassionate presence, it will affect my client through limbic resonance.

> *Ellen is lying on the floor; she finds that if she holds different parts of her body with her hands, it helps her to be more present and tolerate the feelings that arise. As I attune to the kayas it opens a warm, spacious feeling in my heart, I feel a lot of love. Ellen says she feels very safe, she is enjoying being in relationship with her body for the first time.*

Ellen was very affected by the loving and compassionate field we were in together. It is not always so lovely, but if I can be present with what is happening, then it can be both informative and transformational.

> *As Ellen is describing how annoying her husband is, recycling a familiar story, I find myself becoming increasingly irritated and tempted to suggest 'solutions'. As I connect with my inner alignment of the kayas, however, my awareness of sensations increases, opening me to movements or gatherings of energy in my body. A burning heat builds in my chest and I notice the feeling of rage that sits there. I stay with this, rather than react to the energy of it. Gradually she drops into more awareness of her body and begins to feel what is behind the story. As she allows her rage, I feel it less urgently.*

For a long time it seemed that our sessions were the only thing that helped Ellen to manage her suicidal thoughts. Gradually, however, she has built more resources to support herself. Limbic resonance helps me understand how, in our 'relationship loop', initially I was the 'body' that received and helped her to regulate and digest feelings she was unable to bear. This has slowly increased her tolerance so that she has more ability to be present with what is happening and can now connect with physical responses to her emotions. By bringing them into expression through movement, she promotes the flow between the kayas. This meant that instead of being stuck in a story that went nowhere, she could accept the underlying feeling and body experience. This ultimately enabled her to bring more spaciousness to her awareness of the situation.

The Emanation Body

Congruence between the Kayas

The dynamic between the kayas dictates our experience of suffering or freedom. When the mind becomes contracted, through for example, anxiety, grief, or shock, then the flow of our energy body also contracts, resulting in tension, pain and suffering in the physical body. If, however, we can bring more spaciousness to our thoughts and feelings this is reflected in how we embody and express ourselves. Alternatively, when our body is allowed freedom in movement, the process happens in reverse: relaxing our body leads to more flow in our feelings and then more spaciousness in our minds. Although each of us may feel more comfortable in one of the kayas – mind, emotion or body – my observation is that it is the alignment or congruence between all three that enables a flow of energy that promotes health. An experience on the moor showed me this very clearly.

> I am sitting on a hillside in the rain, I am angry! Instead of following the arguments back and forth, I allow the feelings and sensations to fill my body. They are hot and strong, radiating from my heart, into my belly and down my limbs. It is almost unbearable, but as it reaches a terrible crescendo, a new wave comes out from my chest, a clear dark wave that begins to wash through me. Slowly, it begins to settle, the waves keep coming but getting clearer and lighter until finally I am sitting in a luminous, blissful pool of stillness. My body is still here but I experience more emptiness in its nature, it is the calm after the storm. (Autumn 2009)

In this way, by letting go of the story, I allow myself to experience the emotion, the gross aspect of sambhogakaya and I allow it to move through and clarify the energy in my body, moving towards dharmakaya and emptiness. The stronger and more intense the emotion, the clearer I feel afterwards, similar to how it feels after a storm when energy has built up and been released. When I can stay present with what is happening, therefore, it promotes coherence between my mind, emotions and body. This can enable a deep release, not a forced catharsis, but a process that is deeply satisfying and embodied.

Sustainable Practice

Ponlop's term 'unobstructed awareness manifests as body', as a description of nirmanakaya and our ultimate embodiment, invokes for me the fluid and inter-connected nature of our being, described by the model of the three kayas. Ultimately the kayas are three aspects of a whole, therefore not separate or discrete from each other. The trio of mind, emotions and body will be familiar to many; what the model of the kayas brings to this, however, is the awareness of their inherent spaciousness and the flow between them. Through bringing awareness to what is happening for us, in all three aspects, but in particular to sambhogakaya, or the energy body, we open a door for the transformation or movement of stuck energy. This has been known for thousands of years in the Vajrayana tradition, has resonances with other energy medicines such as acupuncture or homeopathy and is echoed in recent understanding in neuroscience and trauma therapy.

Whether it is a body sensation, feeling, or a thought, just as when I move with the wind, remaining present promotes a flow that allows me to process what is happening between myself and my client. Whether it is a soft opening in my heart, the gathering of anger in my chest or even the perception of my own disconnection, through this awareness I begin a process of dis-identification. This means that rather than being in reaction to what is arising, I can allow it to be. It is from this that I begin to experience a quality of presence that does not need to change or interfere with the emerging process. It allows what has been held or fixed to soften and release as I open to the empty and changing nature of what is arising. This in turn potentially opens the space for the client's healing process to unfold.

The support of my relationship to a wider field of awareness and attuning to the flow of the kayas enables greater sustainability in my work. This allows what is arising to move through me, rather than becoming stuck, helping to prevent burnout and giving me more clarity and energy to be present. As I soften my awareness, it opens my heart, not just to my client but also to myself. This helps me to be more congruent with and therefore responsive to my own needs, moving me towards an increasing capacity for love and inner stillness.

Acknowledgements

Thank you, firstly to my husband Rob for his patient help in editing and clarification. Gratitude also to the teachers whose guidance has informed me in writing this chapter, in particular Sandra Reeve, Prapto

Suryodarmo, Ruth Noble, The Karuna Institute, Eliot Cowan and Misha Norland. Also, to my many colleagues and friends with whom I share the journey, to my beloved sons who have taught me so much and for the teaching from the wind, the river, the trees and the land of Dartmoor.

References

Dzogchen Ponlop. (2003) *Wild Awakening.* Shambhala

Kurtz, R. (1990) *Body-Centered Psychotherapy.* LifeRhythm

Levine, P. and Frederick, A. (1997) *Waking the Tiger.* North Atlantic Books

Lewis, T., Amini, F. and Lannon, R. (2000) *A General Theory of Love.* Vintage

Ray, R., Moffett, P., Lee, C., Tenzin Wangyal and Klein, A. (2009) *Start with the Body,* online at https://bit.ly/banda24

Preece, R. (2006) *The Psychology of Buddhist Tantra.* Snow Lion Publications

Sills, M. (2000) *Inner Processes of the Practitioner,* Transcribed from a talk at the Craniosacral Therapy Association, AGM.

Van Der Kolk, B. (2014) *The Body Keeps the Score.* Penguin Random House

Further Reading

Cowan, E. (2014) *Plant Spirit Medicine.* Sounds True

Gerhardt, S. (2004) *Why Love Matters.* Routledge

Levine, P. (2010) *In an Unspoken Voice.* North Atlantic Books

Preece, R. (2014) *Feeling Wisdom.* Shambhala

Anna Murray Preece lives on Dartmoor where the land and river she loves deeply inform and nourish her work. She practices movement healing in the form of Rhythmic Healing, also Core Process Psychotherapy and Plant Spirit Medicine. She runs weekend retreats of movement and meditation to align with the season and explore how our inner sacred essence meets the sacred in nature. She also runs longer meditation and movement retreats in the Vajrayana tradition with her husband Rob and is mother to two lovely young men.

annamurraypreece@gmail.com ~ www.healingfromthesource.co.uk

The Enfolding Body

Embodying the Dynamics of Cell Division

Kim Sargent-Wishart

Abstract

This chapter proposes an experience of the body as a field of awareness, in which continually enfolding cellular surfaces embody the dynamics of reflection, relationship, multiplicity and becoming. Drawing on a Body-Mind Centering approach to embryological and physiological development and embodied practice, my research considers the fundamental biological activity of cell division as 'awareness folding back on itself', as an enactive gesture of one becoming two, and the possibilities for relationship and self-knowing that this gesture allows.

Awareness is understood as an aspect of mind, which permeates all living systems. This is not just about the brain being cognitively aware of the body, but rather the body as an appearance of, and expression of, an aware and unbounded mind, related to the basic ground of being in Buddhist thought.

Awareness permeates all living systems. In the living system of the human body, one of the most fundamental patterns of material organisation is the cell, which not only engages in ongoing processes of respiration and exchange, but also undergoes continual transformation and distribution through the enfolding movements of cell division. The somatic approach of Body-Mind Centering (BMC) defines and locates mind as an intelligent awareness that permeates every level and aspect of our being, and assumes that through guided somatic practice "whatever exists within us as a structure or a function can be embodied and directly experienced" (Haseltine, 2012: 5). Cells have mind, particular tissues or body systems have mind, and when we practise embodying specific structures, tissue or movement patterns, we may experience and express different qualities of mind. The intelligence of the cellular ground – the workings of the cellular collective in configuring our alive physiology – is highlighted as the basis of our bodily knowing, the anchoring and flourishing of awareness in and as a living, biological organism.

Somatisation: Cellular Awareness
Lying down or sitting in a comfortable position, bring your attention to the sensation of breathing. As you breathe, invite the possibility that all of your trillions of cells are also breathing. The exchange taking place between the atmosphere and your lungs is echoed and supported by a deeper, distributed breath exchange taking place through the membrane of every cell. Your cells gently expand and condense with the rhythm of your whole-body breath. Imagine yourself now as one cell, with your skin as an intelligent membrane, an interface between your internal and external environments. As you breathe in and out, there is an exchange of gases and fluids through all of your surfaces, in every direction. You are fluid internally, and swim in a sea of fluid externally. Your membrane determines what substances enter and exit the cell, allowing for flow while providing containment and structure. As this one cell, you gently expand outwards in all directions, and condense in towards centre. There is no place to get to, nothing to be done except to rest in this breath rhythm, equally present to your internal fluid life and the fluid environment you contact and inhabit. Through your membrane, you can meet your surroundings with awareness and equanimity. Enjoy the simplicity of being one cell, breathing.

Cellular breathing and the aware membrane

> Through cellular breathing, we experience life force.
> (Bainbridge Cohen, 2018: 19)

Within the practice of BMC it is common to visualise oneself as a single-celled, amoeba-like organism, to sense breath through all of the body's surfaces, imagining the skin as one's intelligent cellular membrane. Through this membrane, one is invited to explore being simultaneously present to the internal environment, the external environment, and the membrane itself as an organ of transition between inner and outer experience. Cellular Breathing, the exchange of gases through the cell membrane, is one of the primary Basic Neurocellular Patterns (BNPs) of BMC's developmental movement series (Bainbridge Cohen, 2018). Bainbridge Cohen names Cellular Breathing as the "first organic pattern", which "underlies all activity and movement" (ibid: 19). As well as the imaginative experience of being one cell, one can also enter the somatic experience of being a collective of approximately 100 trillion cells, all breathing in a condensing and expanding rhythm that underlies the rhythm of external respiration through the lungs.

Cells are named within BMC as a distributed, collective, 'bottom-up' intelligence; cells differentiate and form communities of function. "Cells are the microcosm of our individual self", manifesting as "both our body and our mind" (Bainbridge Cohen, 2008: 159). Within this cellular paradigm, each cell is aware of stimuli arising from within itself, from neighbouring cells, and from chemical substances and information arriving through the surrounding fluids. Cellular awareness is fundamental to what Bainbridge Cohen describes as the process of embodiment; it is what allows the embodied learning process to occur as more than just the holding and directing of mental imagery from the brain's frontal lobe. While the embodiment process may begin with a cortical image, cellular awareness allows the practice to deepen from a process of visualisation (of anatomical images and descriptions) to one of somatisation, "the process by which the kinesthetic (movement), proprioceptive (position), and tactile (touch) sensory systems inform the body" (ibid: 157). Somatisation invites and fosters a direct experience of cellular awareness, expression and imagination. "To hold an image of the cell in the brain is different than imagining directly via the cells" (ibid: 159). The cortical imagination enters into dialogue with cellular imagination – a more kinesthetic, tactile sense of knowing – and this dialogue informs our cognitive understanding. These two methods of visualisation and somatisation are interwoven in an ongoing developmental practice, in which objective mapping and subjective experience illuminate and reveal each other. The progression through these two methods, though not linear, leads to a third stage, what Bainbridge Cohen calls embodiment, defined as "the cells' awareness of themselves" (ibid: 157), a direct experience of awareness through and as bodily sensation.[1] Bainbridge Cohen further describes embodiment as "a direct experience; there are no intermediary steps or translations... There is the fully known consciousness of the experienced moment initiated from the cells themselves" (ibid: 157).

Linda Hartley, BMC teacher and psychotherapist, portrays the cell membrane as an interface that is both physiological and psychological. She describes this boundary as "awareness—awareness of what is self, what is other, and the quality of relationship between them" (Hartley, 2011: 378). The cell membrane is the interface of communication between the interior life of the cell and the exterior environment. It is active, intelligent, highly flexible, and selectively permeable, choosing what substances pass through. The cell membrane is phospholipid, composed of a double layer of molecules that self-organise through a polar attraction/repulsion to water. Water is the cell's environment. The phospholipid molecules each have a hydrophilic 'head' end, which is drawn to face the inner and outer fluid environments, and a hydrophobic 'tail' end. The tails orient in towards each other, creating the internal structure of the wall, with the hydrophilic heads creating the internal and external surfaces. In a sense, one layer of heads 'faces' the interior world while the other layer 'faces' the exterior world. Echoing Garrett-Brown's (2013) Inter-Subjective Body, we, like our membranes, are permeable, and as such carry the potential for emergent, multiple subjectivities to arise in relationship.

Somatisation: the cell membrane

Imagine yourself as one cell, your skin as your membrane, as you abide in and as a fluid environment. Through your membrane, you are sensitively and actively aware of your environment. Attend to what you sense through surfaces of contact with the floor, the chair, other parts of your body, and with space. Invite yourself to be present to that contact with your environment. This is the fundamental

---

[1] I explore this threefold approach to embodied learning, and its parallels in Buddhist pedagogy, elsewhere. (Sargent-Wishart, 2012).

*movement pattern of yielding, actively meeting whatever you are in
contact with. As you bring awareness to the experience of meeting the
floor, and of meeting the earth through the floor, allow the possibility
of communication to arise through this interface. Yield your weight
to gravity as it meets you, becoming more present, sensitive and alive
to where you are, here and now. Yielding is not about collapsing, but
a priming of the cells in the places of contact, increasing awareness
and presence. How does your sensation of your internal world relate
to this shift in connection to your outer world? Your membranous
surfaces – not too tight, not too loose – breathe, pulsate, and become
more present to the inside, the outside, and what transitions through
in interdependent relationship.*

The cell membrane is not only a lively interface of internal and external
awareness but also always in a state of transition and becoming. As Sandra
Reeve describes the Ecological Body, "like the rest of the natural world, we
are constantly in change or in some kind of transition. In absolute terms,
there is no such thing as a fixed 'position' because there is always
movement" (Reeve, 2011: 48-49). From the moment of our conception,
beginning with the first cell created through the fusion of egg and sperm,
our existence is conditioned by a process of transformational movements,
a complex cellular dance of dividing through dispersal and enfolding. The
first cell, created in a fusion of two, the egg and the sperm, is the *zygote*
(from the Greek 'yoked,' related to the Sanskrit word *yoga*). Our invitation
to corporeality is this yogic fusion of two into one. The tone, or rhythm, of
the zygote establishes a baseline tone that is shared throughout all of a
person's cells, and a basic pulsation pattern of expansion and contraction
begins (Hartley, 2014).

Somatisation: the zygote
*Imagine yourself now as the zygote, this freshly formed unique cell, a
one-celled organism pulsating with possibility, the potential for your
human life. How do you experience the tone, or the mind, of the
zygote? Is there a sense of vibration? Pulsation? Protected deep inside
the cell, within the membrane of the nucleus, lives a new and unique
genetic signature, which will be yours for this lifetime. Imaginatively
sense into the DNA, the genetic strands safely enclosed in the inner
sanctum of the nucleus. This is the signature of your embodiment,
instructions for your embryological development and physiological
activity: those features shared by your species and also those features*

that are unique to you – specific elements of your appearance, preferences and future development. How do you, as the zygote, relate to and communicate with your environment as this unique single entity, in relation to the environment within which you now exist?

Mitosis and cytokinesis: one cell becomes two

The zygote is one cell for only the briefest time[2]; within 12 hours it transforms into two cells, both housed within the same outer shell, the *zona pellucida*. First, the nuclear membrane, within which the DNA resides, dissolves. The strands of DNA separate out of their double-helix configuration, replicate a complete second set, and disperse into the cellular fluid in linked pairs of matching chromosomes. Specialised organelles called centrioles move into position at opposite sides of the cell, establishing a polar orientation. Once at the poles, the centrioles send out tentacle-like spindle fibres that attach at the junction of each chromosome pair, lining up the pairs along the equator of the cell. The pairs then part in a burst of movement from the equator toward the poles, where the two new sets of chromosomes regroup and reorganise into their double helix formation, and the nuclear membranes re-form, completing the first stage of cell division, known as *mitosis*.

Somatisation: mitosis

Imagine yourself again as your first unique cell, your newly created DNA housed within the membrane of the nucleus. As this membrane dissolves, imagine your genetic material, uncontained, unravelling and spreading throughout your internal fluid, replicating itself. Then, sense the centrioles shifting to positions of polar opposition, creating spatial tension. Notice where you sense these poles in relation to your current body organisation: top and bottom, front and back, left and right? Sense the equatorial zone that the poles create between them, along which the pairs of chromosomes align, before, quickly, they part and travel toward the poles and regroup into two identical nuclei. Imagine this brief moment of being one cell with two nuclei, sensing the relationship between them. Does it feel different to sense your

---

[2] As noted by O'Gorman (2013) in her description of the Ontogenetic Body, Bainbridge Cohen claims the first cell is always two cells. While we agree that there is a dynamic of relationship inherent in the story of conception, borne out in the relationship between the nuclei of the sperm and ovum, I feel the generation of the one new cell, before becoming two, is significant.

genetic spark in two places rather than sensing it contained in one?
What possibilities does this being-two provide that didn't exist while
being one?

As these twin nuclei are forming, the next stage of cell division, *cytokinesis*, begins. Now the membrane, that aware interface that maintains the clarity of the cell's inside and outside, actively draws in on itself, enfolding into the cell's spatial equator, until it pinches off to complete cleavage into two genetically identical daughter cells. Proximal, touching, these two cells now share a surface of awareness and communication, created by the enfolding and separation of what was one shared surface. This differentiation introduces connection and relationship, self meeting self in its own reflection, through contact. Inside these two new cells is a distributed identity, borne by identical DNA and other organelles, within shared cellular fluid.

In an essay on doorways and bridges, sociologist Georg Simmel examines the ability to differentiate and separate, and therefore to make connections, noting that "we can only sense those things to be related which we have previously somehow isolated from one another; things must first be separated from one another in order to be together" (Simmel, 1994: 5). To embody the cell we must embody its becoming-two, the continual process of division and differentiation that allows for community and connection.

Somatisation: the membrane enfolds
Again, as one cell, sense the inner layer of the membrane aware of the life inside the cell. Sense the outer layer of the membrane aware of the life outside of the cell. Rest in the balance between the movement of fluids and chemicals passing in and out of the cell, exchanging with the surrounding tissue fluids, and the spatial integrity held by the membrane. Imagine again the sequence of mitosis – the dissolution of the nucleus and replication of your DNA, reorganising into two sister nuclei. Imagine your membrane enfolding and coming into contact with itself, the externally oriented awareness folding inwards, all the way in to centre. What was facing and touching the outer environment now also faces and touches self. An inner surface of knowing awareness is created, as the one cell transforms into two.

Next, imagine each of these two cells undergoing the same transformation, differentiating your wholeness into four sister cells, with multiple surfaces of contact. And then these four replicate and divide, becoming a cluster of eight. With each repetition of the dance of mitosis, you are becoming more differentiated, more complex and layered, with more internal surfaces of contact and awareness. Invite yourself to move as a multi-cellular being. How does it feel to initiate movement from the cells' awareness of themselves, of their community, and of their environment?

From the one-celled zygote, our scale of relationship multiplies, and continues to do so with each subsequent division, creating a tight cluster called the *morula*, increasing in density and differentiation until hatching out of this outer shell into the uterus and implanting into the muscle wall. This primary embryological gesture of becoming two is repeated throughout our lifetime, as cells continue their dance of distribution, recollection, enfolding, and death, in the genesis and continuation of our embodied selves. Our development into a complex organism, with the ability to specialise and grow, even our basic ability to continue living, rests in this early ontological dynamic of shared community and interdependence, with internal communication across aware membranes, among multiple intelligent forms. We are a layered and interdependent matrix of insides and outsides, of dynamically shifting surfaces of connection and communication. The enfolding gesture repeats at other scales throughout embryological development, as evidenced in major events such as gastrulation and the enfolding of the neural tube. The dynamics of our ontological body, "of being becoming and unbecoming, of structures becoming and coming undone" (O'Gorman, 2013: 100) continue to play out as long as we are alive.

Awareness re-flecting back on itself

Membranes as one of the smallest places in our bodies that create structure and contained clarified form....

If we are a one-celled organism, then we can yield with our environment in every direction.

If we are a two-celled organism, then we can yield with our environment in every direction AND we can yield within ourselves, into our own centre.

If we are a 4-celled organism, then we can do as above with more options for knowing ourselves internally and creating multi-dimensional strength and support.

If we are a 90-trillion-celled organism, the yielding options are infinite, and the support of the earth is interwoven into our entire being, and through ongoing cell division continues to be woven anew.

This multiplying of the internal membrane gives structure and form and supports ongoing differentiation and development. So it's not just a matter of making new cells, it's the establishment of a matrix of stability and yielding presence, which allows us to develop as embodied beings.   (research journal entry)

Drawing on Brian Rotman's writing on mathematics and philosophy, I read this process as an animated gesture of re-flection, of awareness folding back on itself:

> To re-flect is to bend back, to fold something onto itself. Folding engenders an ontological novelty, it brings a previously non-existent inside/outside difference into being. In psychic terms, reflection introduces an interiority, an interior space of consciousness and subjectivity. A fold is produced by an animate form touching itself, either via infraceptive (kinesthetic/proprioceptive) monitoring and internal-to-the-brain self-contact, or extraceptively through the myriad gestures of self-touching and self-survey sentient life-forms engage in. In fact, it would seem that consciousness itself comes into being the moment a (mediated) form of awareness—human, animal or otherwise—folds back on itself.   (Rotman, 2009: 76)[3]

In assuming the cell membrane as an interface of awareness, then through the repeated action of cell division, awareness is scaffolded and expanded beyond any apparent dualism of inside/outside or self/other, as complex strata of interpenetrating and co-created multiple insides and outsides with a distributed, communal intelligence. This recalls philosopher Gaston Bachelard's 'spiralled being,' a being that does not have one distinct ('well-

---

[3] What Rotman terms 'infraceptive' monitoring here points to the realm of internally generated stimuli resulting in kinesthetic and proprioceptive sensation, which is more commonly referred to in somatic practice as *interoception*. While the prefix *intero-* refers to interiority, *infra-* means 'below.' *Infraceptive* therefore implies sensation and awareness of what occurs below the surface, and, often below our ability to consciously perceive. It also suggests that this internal monitoring, including cellular awareness, is primary or foundational to ('underneath') exteroceptive experience.

invested') centre, but a distributed, decentralised appearance that incorporates internal and external space into its make-up (Bachelard, 1994: 214). This is not one entity with a concrete inside standing in relief against a vast exteriority, but a multiplicity with countless insides and outsides. In the narrative of embryology as differentiation, distribution, and enfolding, "the dialectics of inside and outside multiply with countless diversified nuances" (ibid: 216) which, furthermore, continue to appear and disappear in every moment of embodied life.

When I engage with this narrative of my body as trillions of enfolding, replicating cells while moving, I experience a sense of infinite articulation of multiple sites of cohesive intelligence.

When I engage with the dynamics of distribution embodied by my ever-dividing cells I become more somatically and cellularly aware and activated.

When I embody the folding-in process I meet new surfaces of my immediate environment and my own interior landscape. I am surprised by new surfaces becoming available, dialogues that open up through new contingencies. I discover my body-mind as an intelligent system always creating itself anew and blooming with possibilities.

Differentiation and wholeness: body as community

> What do we know of ourselves through the inner touch of our trillions of surfaces, which are continually spawning, swarming, specialising into organ and tissue, and sloughing off? What intelligence, like the optimal pace of a heartbeat, is shared collectively between and among them? (research journal entry)

Noting that cells rely on the functions of internal respiration and "awareness of internal and external stimuli" for their survival, Bainbridge Cohen writes that, additionally, "in order for a community of cells to survive, they also need to have an interconnecting feedback mechanism from each cell to all cells. Hence, each cell has a sense of self and communicates with all the other cells" (Bainbridge Cohen, 2018: 21). The coherent wholeness of a body is continuous across all of its differentiations. Embodied life, as a relational and material field of awareness, coheres through the transformative movements of differentiation, distribution, and enfolding enacted in the creation and replication of every cell. Differentiation through cell division creates individuation and community simultaneously. Through the yielding relationships that occur at and across cell membranes, and the shared interior cellular substances, community creates the organism.

Embryology by definition is a narrative of creativity, of the play of life force through the emergent materiality of the body-mind. From a view of primary wholeness, such as the Buddhist notion of an undifferentiated, intelligent ground of being (Guenther, 1987), or the Tao, the "mother of all things, …invisible and unfathomable" (Chang 2011), or the 'holo-movement' described by physicist David Bohm (2012), the creative appearance of form is frequently described as a process of differentiation of an ultimately nondifferentiated background oneness in a play of becoming aware of itself; an impermanent display of materiality which is interdependent with everything else in the universe. The enfolding of the cell membrane can be understood thus as an embodiment of the generative process of differentiation within a larger field of awareness, mind, or being. As Chuang Tzu, one of the founders of Taoism, said "…all things create themselves from their own inward reflection and none can tell how they come to do so" (*Chuang Tzu Ch. VIII* in Chang, 2011: 92). In this way, awareness becomes aware of itself through embodiment – the creative, generative and mysterious process of living.

References

Bachelard, G. (1958/1994) *The Poetics of Space*, tr. M. Jolas. Beacon Press

Bainbridge Cohen, B. (2008) *Sensing, Feeling and Action: The Experiential Anatomy of Body-Mind Centering*, (2nd ed). Contact Editions

_____ (2018) *Basic Neurocellular Patterns: Exploring Developmental Movement*. Burchfield Rose

Bohm, D. (2012) *On Creativity*. Taylor and Francis

Chang, C. (2011) *Creativity and Taoism: A Study of Chinese Philosophy, Art and Poetry*. Singing Dragon/Jessica Kingsley

Garrett-Brown, N. (2013) 'The Inter-Subjective Body' in S. Reeve (ed.) *Body and Performance*. Triarchy Press

Guenther, H.V. (1987) *The Creative Vision: The Symbolic Recreation of the World According to the Tibetan Buddhist Tradition of Tantric Visualization Otherwise Known as the Developing Phase*. Lotsawa

Hartley, L. (2011) 'Boundaries, Defense, and War: What can we learn from embodiment practices?' in G. Wright Miller, P. Ethridge & K. Tarlow Morgan (eds.), *Exploring Body-Mind Centering: An Anthology of Experience and Method,* North Atlantic Books, pp. 377-390

_____ (2014) 'Embodiment of Spirit: From embryology to Authentic Movement as embodied relational spiritual practice' in A. Williamson, G. Batson, S. Whately & R. Weber (eds.), *Dance, Somatics and Spiritualities: Contemporary Sacred Narratives.* Intellect, pp. 9-34

Haseltine, R. (2012) 'Holding the Whole: BMC concepts and principles', *Currents: A Journal of the Body-Mind Centering Association,* 15(1), 4-7

O'Gorman, R. (2013) 'The Ontological Body' in S. Reeve (ed.) *Body and Performance.* Triarchy Press

McGuire, M. (2011) 'Body-Mind Centering, Mindfulness, and the Body Politic' in G. Wright Miller, P. Ethridge & K. Tarlow Morgan (eds.), *Exploring Body-Mind Centering: An Anthology of Experience and Method.* North Atlantic Books, pp. 369-376

Reeve, S. (2011) *Nine Ways of Seeing a Body.* Triarchy Press

Rotman, B. (2009) 'Gesture and the 'I' fold', *Parallax,* 15(4), 68-82

Sargent-Wishart, K. (2012) 'Embodying the Dynamics of the Five Elements: A practice dialogue between Body-Mind Centering and Tibetan Buddhist philosophy' *Journal of Dance and Somatic Practices,* 4(1), 125-142

_____ (2016) 'Making Nothing out of Something: Emptiness, embodiment, and creative activity', PhD thesis, Victoria University

Simmel, G. (1994) 'Bridge and Door', *Theory, Culture & Society,* 11(1), 5-10

Kim Sargent-Wishart PhD is an artist, researcher, writer and educator specialising in somatic movement methodologies and creative practice. Her research and practice interests include embryology as a model of creative process, somatic methodologies for practice-led research, dance and screendance, contemplative photography and somatic writing. Kim is a certified practitioner and teacher of Body-Mind Centering, and the Administrative Director of Somatic Education Australasia. She lives on the Bellarine Peninsula (Wadawurrung country) in Australia.

kimsargentwishart@gmail.com ~ www.kimsargentwishart.com

The Imaginary Body

In the relationship between body and awareness, what is the role of imagination?

Alex Crowe

Abstract

Interwoven with a description of a sequence of movement experiences, this chapter is an exploration of the role of the imagination in the relationship between body and awareness, using ideas from Kantian and post-Kantian philosophy about imagination.

Defining Awareness

For a working definition of 'awareness', we could begin with what in the Buddhist tradition is called 'sati-sampajana', or "alert but equanimous observation" ('sati') combined with "the ability to fully grasp or comprehend what is taking place" ('sampajana') (Analayo, 2003: 60). In this definition, awareness has two aspects. Firstly, it involves the conscious registering of present-moment sense data (i.e. experience appearing in the modes of the five physical senses) and present-moment mental activity. Secondly, it involves recognising, in some way, the nature of what is manifesting in experience. In this respect, what we are calling 'awareness' shades into what might often be labelled as 'perception', taken to include a process in which sense-data become organised, or more organised, into a gestalt of some kind, rather than as a more amorphous wash of colour, sound, smell, etc. (We leave aside here the question of whether there is such a thing as entirely unorganised, completely 'raw' sense data.) This organisation of sense-data takes place within the process of perception, rather than manifesting as independently identifiable mental content. Thus, if I look at or feel my hand, I experience *my hand*, rather than experiencing various sense data and thinking of them, or picturing them in my mind's eye, as 'my hand'.

> Loosely spreadeagled on the living room floor, I'm daydreaming about something-I-know-not-what. And then, suddenly, I'm here. "Oh yes: movement practice." Sensations of floor pressing into tissues, a sense of the shape of this posture. Moving, I am soon on hands and knees, feeling weight shifting, muscles working and releasing, the sense of the solidity of the body and its structure, a sense of the bones. I notice the table nearby, its structure and nature seeming analogous to my own. The whole room resolves into a solid, structured space of real-seeming, solid objects, and I too feel more real. Thoughts continue to pass through, but my awareness and interest is anchored more in physicality, and they do not, for the moment, take me away from here.

The body in awareness: the role of imagination

Let's apply this definition of awareness to generate a phenomenological definition of 'body', i.e. body-in-awareness. It involves, firstly, present-moment awareness of what is manifesting in the haptic sensory mode; and also, secondly, the appearance of those haptic sense data as a gestalt. Body-in-awareness, in other words, is the experience of a particular mapping of the field of haptic sense-data.

This phenomenological definition is not what we normally mean by the word 'body'. Normally we mean the familiar material object with arms, legs, teeth, kidneys, etc. Thus in the movement description above, the initial coming into the here and now, and into the body-in-awareness, was an experience that was strongly inflected by my familiar, habitual mapping of haptic sense-data, which involves knowledge about normal human physical shape and anatomical constitution. But, as any somatic practitioner or meditator knows, the body-in-awareness can be very different from how that familiar material object appears from the outside. If I pay closer attention to my haptic sensations while lying on the floor than I did in this particular instance, perhaps in a more meditative frame of mind, I cannot actually distinguish, in the sensations themselves, physical body and floor; there are just these sensations of hardness, pressure, temperature. It is a common enough experience that body extensions and prostheses readily get mapped into, included in, body-in-awareness – walking sticks, violins, tennis racquets, even cars, become at least as much part of it as kidneys and teeth. .

Moreover, and significantly for the purposes of this chapter, the haptic sense is not isolated in experience from the other physical senses, so that experience in other sensory modes can also be mapped into the body-in-awareness. For example, in many somatic practices, visual imagery is used to stimulate particular experiences of the body-in-awareness. Thus in Skinner Releasing Technique, one may experience a body made of spaces, silk or mossy rainforest. Such experiences may begin merely with awareness of a mental image and also, separately, of one's more familiar, anatomically-based haptic mapping of the body. But they can go beyond this to an experience the body merging with the image, so that the body-in-awareness actually *becomes* (say) a rainforest-body. Sound qualities of music can also be experienced in this way, not only as *in* the body but *as* the body – 'melting' music merging, in experience, with melting tissues. Or if one pays close attention, how does the fragrance of honeysuckle or the taste of mint affect the experiential gestalt of the whole body-in-awareness? Are they not, to some extent, incorporated into it?

Such phenomena are presumably related to synaesthesia, though a further discussion of that question is beyond the scope of this chapter. In any event, the effects of overtly poetic imagery in somatic practices show how the body-in-awareness involves not only an organisation of haptic

The Imaginary Body

sensation but also of haptic sensation merged with *mental representations* (i.e. imaginings) of experience in other sense modalities – most commonly the visual. And this sheds light on how more prosaic experiences of the body-in-awareness, based on scientifically derived anatomical imagery, also involve imagination, whether they occur in somatic practices that deliberately deploy anatomical imagery, or in everyday life. Body-in-awareness is always inescapably imaginary.

> *I'm on my feet now, enjoying the sensations of my physical structure moving among the structures of the space around me – bones and tissues in shifting alignments, weight and breath. I am quite present, but the movement passes away as it arises, and I will later retain no memory of it – until a moment when I suddenly realise I am edging my way towards the corner of the room by a window, where an open curtain is gathered in a long column of fabric. Suddenly my present experience becomes sharper and more vivid. It feels inexplicably significant to me that I am approaching this curtain, attempting to get closer to it. My arms are outstretched to each side, and my chest area feels open, yet my sternum area feels hard, and I experience it as made of tin can material. My torso no longer feels as if it has its usual volume but instead feels flat, as if two dimensional. My whole body is made of thin pieces of a white, brittle substance, like dried up cuttlefish. My legs are held close in a narrow stance, with the feet slightly crossed, and feel unusually long, like stork's legs, and they seem stiff and unbending. With legs like this I cannot step out and forward very far, and as I edge along I feel how tentative, how difficult, this approach to the curtain seems to be.*

But it may seem puzzling that body-in-awareness, based as it is in haptic sensation, is imaginary. Awareness of sensation feels, in my experience, like contact with reality. Compared to being lost in thought, in some fantasy world of my mind's creation, feeling my body can give a feeling of relief, of safety and stability, of knowing where I (really) am, who I'm (really) with, even who I (really) am. This coming back to the reality of the here and now is the stock in trade of discourse around mindfulness. And awareness, as defined above, involves true perception of reality. But how can a body made of cuttlefish be called 'real'?

The puzzle here arises from the binary opposition of 'real' and 'imaginary' – an opposition which is indeed legitimate in everyday usage ("I thought I heard the doorbell, but I was only imagining it"), but involves problematic metaphysical assumptions. 'Real' can be taken to imply the existence of an objective world, probably a material one, independent of the experiencing subject; and then is readily associated with haptic experience of physical objects, including the body, and thereby with sensation in general. 'Imaginary' is associated with the mental activity of the experiencing subject, conceived as categorically separate from the objective 'real' world. On this view, sensation gives access to the real (external) world, imagination to the merely invented internal world of mental images.

But what is imagination?

The philosophical, phenomenological and, since the 20th century, empirical difficulties with this kind of naive realism are well known. Yet, in my own experience, it is extraordinary how tenacious a delusion it is, even if one is aware of its deficiencies. Extraordinary, that is, until one remembers that so venerable a tradition as Buddhism describes it as the fundamental delusion of human existence, one that it usually takes more than a lifetime to see through. Experiential enquiry into the role of imagination in body and awareness may be a useful way of engaging with this delusion, and to support that enquiry it may be helpful to turn to philosophical analysis of the concept of imagination (Warnock, 1976).

For example, the philosopher David Hume describes how he thinks the imagination is involved in the process of what I have called awareness, in recognising the objects before me for what they are; and more generally in the construction of the world as we perceive it. German philosopher Immanuel Kant's more elaborate theory distinguishes imagination from the faculty of 'understanding'. 'Understanding' is the part of us that marshals fundamental categories that structure all our experience (such as time, space and causality), and other concepts dependent on them. Imagination, by contrast, is the faculty which maps immediate sensory experience, which would otherwise be incomprehensible chaos, onto the concepts with which we categorise our world. Concepts themselves are like empty vessels, potential forms in and through which experience might be comprehended but, in themselves, not part of experience. It takes imagination to bring them to life, to fill them with raw sensory experience in generating perceptual experience.

What emerges from accounts like these is that imagination has a crucial role in giving form to experience – in other words, they explain how imagination is part of the process of perception, and thus how imagination is built into awareness (in our definition of the latter). Imagination is that aspect of consciousness that finds (or creates) patterns in experience, and so presents us with a world that is graspable, intelligible. And in such accounts, contrary to the naive real-imaginary binary, it is the same faculty of imagination that is at work not only in ordinary perception of what we take to be the external world but also, equally, in creative process, where perceptual experience changes and the everyday conception of the world is transcended such that 'reality' is experienced in a new, unfamiliar way.

We can use Kant's ideas on imagination to illustrate this. In ordinary perception, as we have seen, imagination fills out our familiar concepts with the sense data at hand to generate the gestalts of our world as we habitually experience it. In creative process (i.e. in relation to what he calls 'aesthetic' experience), the perceptual process changes, in that the imagination finds its own, unique ways of organising sense data, based on order inherent in them that does not correspond to any of our existing concepts. The result is an experience of the world that is fresh, has a sense of freedom about it, and that, for Kant, is constitutive of beauty. I find it interesting to interpret the movement sequence above in these terms. An initially rather mundane, forgettable series of perceptions suddenly becomes vivid and alive, and I experience my body in unfamiliar, even bizarre, ways. One could say, perhaps, that this is the imagination starting to function 'aesthetically' as part of creative process – no longer finding form in experience on the basis of established concepts, but beginning to present the world (and the self) in a fresh way.

Imagination as a faculty of feelingful knowing

I come to a halt about a foot from the curtain and seem to be able to go no further. My left hand reaches slowly out towards it on the end of a rigid arm which moves, at the shoulder, like a rusty old door closing. The fabric is dry and rough to the touch, and I am surprised; I had, unwittingly, been expecting something silky and lush. My mind is blank. Then, suddenly, apparently out of nowhere, the thought comes up in my mind, with a half-formed image, of someone I have loved and lost, and there are tears flowing down my cheeks, and my heart is welling with grief.

Kant's theory, variously interpreted, underpins notions of the imagination in German post-Kantian transcendental idealism, and thence has influence on those of continental phenomenological philosophy and, via

Samuel Taylor Coleridge, those of the English romantic poets (Hill, 1977). We can turn to Coleridge for further illustration on how, in this tradition of thinking, the imaginary is not opposed to the real but (at least partly) constitutive of it, not only in the basic unconscious process of everyday perception, but also in conscious creative process – how the imagination, in other words, is a faculty of knowing. Coleridge also emphasises the relationship between imagination and feeling, which I find helpful in interpreting experiences like the one just described.

For Coleridge, an important aspect of imagination working creatively is that it deals in symbols. It does this primarily in perception, so that objects appear to us not (only) as what they are according to our usual conceptual mappings, but also as symbolic of something else. Indeed we may experience them not merely as pointing to or representing that something else, but even as *being* it, acting as a living metaphor for it (as when experiencing one's legs as stork's legs, for example). Secondarily, one may produce a representation of this symbol-in-perception (drawing a stork-legged person, for example) – and this, in Kant and Coleridge, is the role of the artist.

It will be evident that imagination, working creatively in this way, generates awareness, in the sense defined above – awareness that is both a conscious registering of the present moment sense data; and also a grasping of the sense data as a gestalt, an understanding of the whole situation that is not merely abstract and conceptual, but an immediate, sensory experience. As Coleridge emphasised, there is an aliveness in such awareness, both in the sense that it transcends our habitual conceptual framework and, given its foundation in sensory experience, that it is integrated with feeling. He described imagination, working in this way, as a unifying, "combining power" (ibid: 78), conjoining the understanding of a general truth of some kind with the particular experience of the present moment, yielding a grasp of something about life that ramifies beyond, but is also contained in, the here and now: "the union of deep feeling with profound thought" (ibid: 101).

The sublime: imagination as a way of knowing the unthinkable

Language like this may have some bearing on the experience of strong emotion in movement, as in the instance described above. But there is a strong strain of mysticism in Coleridge, and his talk of the 'deep' and 'profound' also points to another aspect of imagination, and another aspect of imaginative knowing. To unpack this, it is helpful to return to

Kant. In Kant's account of imagination, there are two distinct ways in which the imagination presents us with experiences that are beyond the reach of our usual, conceptually-based perception. Firstly, the imaginative experience could be of something for which one already has a familiar gestalt, based on an existing concept. As we have seen, any imaginative experiencing of the body would be an example of this, given that we have a well-worn set of concepts for the body corresponding to ordinary ways of perceiving it, and that experiences of body made of fluids, silk, cuttlefish etc., would subvert. Working in this way, we can gain a vivid experience, a particular kind of knowledge, of something that we may otherwise understand merely intellectually without finding a counterpart for it in actual experience ("I will never again be able to touch this person who has gone"; or "Everything is impermanent").

But secondly, and more radically, the imagination could also provide a way of experiencing something which has not been mapped, or even *cannot* be mapped, using our usual conceptual system. This is Kant's theory of what he calls 'the sublime'.

> Slowly I move back from the curtain, moving more freely, my body feeling now more fleshy and familiar, breath swelling the torso, emotions settling and grounding. A sense of energy rising in the body leads me into arching back, arms reaching into the space above me. I catch sight of the moulded ceiling rose above me and say the word 'rose', and stumble back to plump down in an armchair directly opposite the curtain. On the table next to me is a china bowl of rose petals that belonged to the person I have lost, and on an impulse I take it in my hands, holding it delicately at my fingertips, rise and move on soft knees through the room, with a feeling of gentleness and poise. At a certain point, I find myself slowly, carefully, placing the bowl on the floor before the curtain, though several feet away from it. It feels like making an offering, though to what or whom I do not know – not, I feel, to the curtain or whatever it might represent to my embodied psyche. The movement is light, with mindfulness but without ceremony, and when I come to standing again I know something has finished.
>
> I return to the armchair, and notice how it feels as if the whole space is alive; and that it seems as if the area bounded by the ceiling rose, the curtain and the bowl on the floor is particularly dense and vibrant. In my mind's eye, I see delicate lines connecting them, and the triangular space between the lines very slightly shimmering. It is

an uncanny feeling, and there is a sense of significance; though it
seems clear to me that that significance is nothing to do with my own
personal history and my emotional response to it (which has passed),
or at least not only to do with that.

One can, perhaps, interpret this movement experience in terms of Kant's ideas about the sublime. He argues that the human mind can operate with concepts for which no corresponding experience is possible because they are so abstract ('ideas of Reason' in his terminology) – such as those of freedom, or mathematical concepts such as very large numbers (Warnock, 1976: 65); or perhaps, one might say, of infinity, of God or of consciousness free of any subject-object division. Without going into the minutiae of Kant's theory, we can perhaps take it quite broadly as pointing to the possibility that there are ideas we can think about, thoughts we can formulate conceptually and that make sense logically, but for which there can be no counterpart in our experience other than symbolically (perhaps such as "emptiness and form are not different").

I take it that symbolic experiences of this latter kind of idea correspond to what Kant calls 'sublime' experiences of 'aesthetic ideas'. They are accompanied by feelings not only of aesthetic appreciation, but more of awe, and a sense of mystery. Such feelings may be something of what Coleridge had in mind when he wrote of how the imagination can create an 'atmosphere' of what he calls 'the ideal world', which affects "forms, incidents and situations, of which for the common view, custom had bedimmed all the lustre, had dried up the sparkle and the dew drops" (ibid: 101). Thus the imagination is a 'magical power' (ibid: 93), with the ability to transform the familiar, mundane world.

Conclusion

So it is that the creative imagination, one could say, re-enchants the world. Its relegation in favour of more literalistic conceptual, functionally oriented thinking has been an important strand in critiques of modernity from the romantic poets through to contemporary accounts of the imagination such as the writing of Patrick Harpur (2002). The body, too, has a central role in such critiques, one which often begins with the consequences of abandoning a phenomenological perspective, abandoning body-in-awareness, in favour of a scientific one which assumes the primary reality of the 'objective body' (Reeve, 2011). And this is no coincidence if, as is my own experience and a common theme in somatic practices, awareness of the body is a gateway to the creative imagination. Awareness of immediate sensory experience loosens the hold of our conventional mappings not only of body but of the world as we perceive it, while also, in doing so, providing contact with something real and indubitable, and thereby offering what could be called a direct form of knowledge. Kant and his philosophical successors offer insights into how that loosening can be seen as a shift in the functioning of the imagination from its concept-bound mode in ordinary perception to a more creative mode – a mode in which we grasp experience in terms of symbol and metaphor, often integrated with feeling; and which provides a way of knowing that transcends our concepts and our language, and perhaps, in the sublime, transcends any possible concepts at all. In inviting, interrogating and reformulating that direct knowledge moment by moment, moving with awareness, then, can indeed be described by dance scholar and philosopher Maxine Sheets-Johnstone as "thinking in movement" (Sheets-Johnstone, 2011: 419) – a form of 'thinking' that essentially involves the creative imagination.

References

Analayo (2003) *Satipatthana: The Direct Path to Realisation*. Windhorse

Harpur, P. (2002) *The Philosopher's Secret Fire: A History of the Imagination*. The Squeeze Press

Hill, J.S. (ed.) (1977) *The Romantic Imagination: A Casebook*. Macmillan

Reeve, S (2011), *Nine Ways of Seeing a Body*. Triarchy Press

Sheets-Johnstone, M. (2011) *The Primacy of Movement* (2nd ed). John Benjamin

Warnock, M. (1976) *Imagination*. University of California Press

---

Alex Crowe is a mover, performer and performance maker. Skinner Releasing Technique, of which he qualified as a teacher in 2006, is a key influence on his practice, as is the work of Suprapto Suryodarmo. He holds an MA in Performance Making from Goldsmiths College, and teaches movement, improvisation and performance in both Higher Education and community settings. He has been practising Buddhism and Buddhist meditation since 1997 and currently works at the West London Buddhist Centre, where he has led workshops and offered performances that seek to integrate Buddhist and somatic movement practices.

aewcrowe@yahoo.co.uk

The Instrumental Body

Olga Masleinnikova

Abstract

Drawing on choreology I share my observations on possibilities offered by movement analysis, to initiate and expand sensorial awareness. I introduce the notions of the *Home Body* and the *Instrumental Body* which provide an ON/OFF button as an intentional tracking system for sensory experiences within performance practice.

Using a practical session with actors as a case study, I present a process where an intentional shift into the attitude of *Instrumental Body* enables movement tools to become an awareness map for the transformation of patterns, for the experience of new inner landscapes and for creative expansion.

Everything is a matter of balance

This is my first session with a collective of ten young professional actors. I am invited by the director V. to share some practical movement tools to create both stylised movement and quick naturalistic physical transformations so that they can play multiple characters in the same show. I ask them to walk around the space – a bright, airy studio with large windows overlooking surrounding trees. Each one of them walks with a slightly different quality: some are faster, others are slightly heavier and slower, one glances outside, another looks at the floor. After suggesting that they stop, I ask the group to share what they remember from this walk. G. says that he noticed that he slowed down with time; L. enjoyed seeing the trees which made her feel more peaceful as she was a bit nervous at the beginning of the session; M. added that he felt the same but he enjoyed hearing the sounds of the birds outside.

V. says: "We all swung our arms". *They all nod their heads. I ask them to walk again in their most comfortable way and to pay specific attention to the swing of their arms. When we stop, we discuss their observations. They realise this functional arm swing is very different for each one of them. G's swing is slightly bigger in size, M's swing is very tiny, and her arms are very close to her hips, J's right arm swing is heavier than his left arm swing. We play a game. I ask R. to demonstrate his arm swing, they copy it. All their attention turns to R, they do their best to imitate him. I ask them how they felt. They share:* "It is interesting how focused I became"; *another:* "Yes, my entire attention was on R". *This time I turn to V, but instead of performing her arm movement, I ask her to describe in words her physical experience. She says:* "Right arm light with a small swing, left arm turned slightly inwards, slightly bigger in size than the right". *I ask the group to walk but to use their arms as described by V. They try and I ask them how they feel. The words are* "interesting", "I feel so different", "I never thought of my arms like that before". *This time I ask G. to describe the movement of his arms. He says:* "fast, with a tiny flick outwards on the right side, left arm is slightly heavier". *We try to reproduce his unique style. This time the comments are,* "it is so different", "I love that little flick", "I feel playful", "I feel like I want to get somewhere fast", "it is hard to flick only on the right side", "what is interesting is that my whole body changed, I started walking faster". *I ask them to regain their usual arm swing, as well as their comfortable speed. They breathe out, I hear,* "oh it feels nice". *L. adds* "yes, I feel like I just returned home". *I ask them how this was different from copying R's movement; they share:* "my attention is within myself", "I really felt my body changing". *I conclude this task by introducing the concept of organic order that makes our unique movement pattern feel so effortless.*

Rudolf Laban observed human movement and identified a clear structure of patterns and tendencies. He then assembled his findings into a system of analysis to demonstrate that human movement is highly organised. Choreology, devised by Laban's student Dr Valerie Preston-Dunlop, is a contemporary development of Laban's principles, specifically for performing arts. She assembled some of Laban's work in a 'structural model of human movement', that identifies five visible and simultaneous components of movement, naming and differentiating body, action, space, dynamics and relationships. Each of these components is also a complex simultaneous organisation of patterns that tend to relate to each other in a systematic way, allowing human movement to follow a particular sequence, some sort of order, a hidden grammar that makes movement feel effortless and natural. Laban explained it as "choreological order through which movement becomes penetrable, meaningful and understandable." (Laban, 1966: 5)

Preston-Dunlop writes: "Laban asserted that the factors of motion, its weight/force content, its spatial form, its timing and its flow content are signifiers in all human performance" (Preston-Dunlop *et al.*, 2002: 66). Because of these movement factors and its resistance to the pull of gravity, human movement is based on opposition in order to keep a balance. Each movement needs a certain amount of tension and release in our muscles to resist the force of gravity. This enables limbs to swing forwards and backwards, it allows us to flex and stretch when we need to grab something, the torso to open and close when we breathe, to transfer the weight from one foot to another, in an efficient, non-stressful manner. The energetic balance between tension and release is essential to keep movement efficiency. With too much relaxation we fall to the ground, with too much tension we freeze.

"Dynamic balance arises through the connection of one movement to the next, through the natural preparation-action-recovery sequence of movement. One move acts as the preparation for the next. Its recovery becomes, in its turn, the preparation for the third, and so on" (Preston-

Dunlop, 1998: 106). This order is essential to move efficiently, and it takes place sequentially and simultaneously during motion. Choreologist Rosemary Brandt says, "Choreological order is the term Laban gives to the syntax of human movement" or "the choreological order is what holds our movement together" (Brandt in Preston-Dunlop *et al.*, 2002: 64), "we don't have to think about it, we haven't been taught how to do it, we do it because it feels natural and comfortable" (Brandt, 2008).

I like to call this being in the *Home Body* mode: following the order, in a subjective way. *Home Body* or the body in which we live, the mode from which we function and accomplish daily tasks. For example, walking from A to B: this tends to happen forwards, in an upward position, our feet generally unfolding from heels to toes, preferably in a straight line with a lightly curved pathway when/if we change direction. If, during our walk, we drop something onto the floor, generally we would bend forward and use the internal surface of our hands to grab it. Another example would be our need for recovery, our sleep, ideally in a comfortable, safe space that allows us to lie down, if possible, in our favourite position.

As demonstrated in the first task, all these effective changes happen slightly differently for each one of us. These are just examples of our functional, human movement patterns informed by our body structure/cultural/emotional/genetic/gender/health and other systems that unfold in an organic order, following a cycle, keeping the balance and simultaneously making our movement unique and recognisable. In the *Home Body* mode, we don't have to make a specific effort to think about how we are going to use our bodies; we use them in an efficient way in order to be able to function. *Home Body* is a functional container for our recognisable signature in the world.

That same idea of order, of patterns, exists in nature, something unseen, yet ever present; the cyclical manner in which natural phenomena occur and the simultaneous force that emanates from, and unites, all the different expressions of life . The concept of choreological or organic order in human movement could be compared to this natural organisation. A law of harmony that governs in nature, as well as in our body and our movement.

Same body, change of intention

I ask all the participants to walk around the space but this time I ask them to engage with their body as if it were their instrument. Like piano notes, our body parts and surfaces can be used as notes within our Instrumental Body. I suggest that they train their proprioception beyond the usual comfort zone of their Home Body, and that they expand their awareness to less evident surfaces and body parts. I invite them to transform, intentionally, their usual way of placing their feet when they touch the floor, by shifting the weight to a different surface. I ask them to have a lighter step, to change the tension level in their chest, or to transform slightly the size of the movement or the opening of their eyes. By varying the use of space, time, flow, force, weight and by introducing new possibilities to body parts and surfaces, we are creating the layers of a new physical experience and awareness. I am witnessing physical changes, quick transformations and clarity of physical details.

This time I ask the group to be active players by making their choices independently. I invite them to try small changes in their usual movement signature by playing with movement tools such as tension levels, speed variations, subtle weight transfers or slight size modifications. By asking the group to make minor changes in their pedestrian walk, they transform but still remain within the naturalistic style. From there I begin to intentionally direct the movement to create details by emphasising some surfaces or body parts, transforming the way they touch, adding repetitive gestures or working on their head nods, changing the way they breathe, or how they use their voice and eyes.

To end this part, I ask them to shake their bodies to release the accumulated tension from the physical changes and invite them to reflect on their process, to integrate their experience by anchoring their personal preferences in writing, to record what was more or less comfortable. I then ask them to think about their characters in the performance by selecting a series of physical anchors for each. I remind them to choose what would work best for the character, it could be comfortable or similar to their Home Body, or it could be challenging. I ask them to include variations of speed, use of personal space, tension levels and distribution of weight, to create a clear score by drawing or writing. These could be used later in the creative process as a visual stimulus, as process documentation and as visible traces of character development.

I was asked to give these actors tools to create naturalistic physical transformations. As a point of departure, I chose a pedestrian walk and addressed body parts as notes of the *Instrumental Body*, creating a dynamic expressiveness that felt corporeally different to the mover, but remained naturalistic in style. A musical instrument like a piano, for example, has been created by a human knowing how it works. A human being is an alive process and is not delivered with a 'how to use me' leaflet. Our body and its systems are supposed to move harmoniously from our inception and would still function without us understanding how it all works. In this case, the actors' acquired awareness, first of their *Home Body* preferences, followed by the intentional changes of their body as an instrument, has become a tool for the desired transformation of physicality and style.

Nature, body and human movement do not work in isolation, therefore by changing one part intentionally, the rest of the body tends to adapt, to recover, to create a harmonious, coherent whole. Even if it feels unfamiliar to the subject's body, less natural because of the habitual pattern, the body will tend to adapt itself to follow the organic order, as everything is a matter of balance. Dynamic expressiveness is achieved through an alternation of contraction and release.

For example, when the mover is experiencing a bound action, which demands a certain amount of tension in his muscles, he will recover from this tension by producing a free quality movement somewhere else in his body or will be followed by a movement that will allow the necessary release. By counterbalancing one dynamic with its opposite through the contraction or release of energy, our body can find a sequential or simultaneous balance in motion. If we look at a slow, sustained movement, it will tend to be followed by a fast, sudden action. A strong gesture will tend to be balanced by a softer quality movement. A direct movement in space would tend to be followed by a flexible movement in space. Besides, each dynamic has a simultaneous 'tendency of direction' in space and a simultaneous 'tendency of speed'. For example, a light quality will tend to go upwards, and a heavy quality would tend to go downwards. Each action[1], such as jump, close or open also has an inherent rhythm or speed that will occur when producing that action organically and efficiently. A fall is visible if it is executed with a direct sudden quality to it, or a functional, relaxed body opening tends to use a deceleration of speed.

---

[1] Choreological Studies identifies 11 actions in the structural model of human movement: travel, isolation, transfer of weight, turn, twist, jump, fall, open, close, stillness, lean.

Additionally, a human body is three dimensional and because of this structure obeying the laws of gravity, time and space, the mover manages efficiently her/his energy flow, weight and force[2] by following a particular sequence. For example, one move to the left side tends to be balanced by a move to the right. A motion forwards tends to be balanced by motion backwards and a movement upwards inclines to be balanced by a motion downwards. Straight shapes will tend to be followed by curved ones and moments of stillness are necessary to recover from motion. We need this sequential balance to function in an efficient and healthy way, to follow the law of harmony.

'Break the balance' games

Game 1: Breaking the simultaneous order of movement

I make the task more complex by activating several parts of the body simultaneously. This time I decide not to follow the dynamic balance and take away the function of simultaneous recovery, by giving a specific indication to the body part that is supposed to recover from the physical change. I give them an indication for the right foot, followed by a different one for the left foot; I add a movement of the chest and conclude by a contrasting movement of the eyes. Instantly this body chord heightens their proprioception as well as their mental effort, they are fully present, trying very hard to 'play' the expression. I ask half of the group to stop and watch the other half. They say: "they look robotic", "weird", "disturbed". *I ask them all how they felt while doing it:* "It was very hard, much harder than before, I had no idea who I was or how I felt, I couldn't connect the dots within myself".

Breaking dynamic balance by using a polyrhythmic chord within a pedestrian context created a very different sensorial awareness and a less 'natural looking' or 'inorganic' visible outcome. By choosing to use movement tools to purposefully guide the body into an expansion of preferences and functional systems, I aimed to create an embodied understanding of how movement works, a visceral sense of the ever-changing flow and an experience of the stylistic shift, but from the corporeal perspective. The result obtained from these changes created a

---

[2] Rosemary Brandt developed Rudolf Laban's Effort Graph by adding the motion factor of Force.

different phenomenon: it still had the traits of human movement but perhaps more robotic, artificial, or something more abstract, like a feeling of something that needed to be fixed. By breaking the dynamic balance, we left behind the naturalistic style.

Game 2: Breaking the sequence of movement

For the next task, instead of functionally placing our feet from heel to toe while walking forwards in a daily context, I ask them to alter this sequence to toe to heel. This new foot pathway created a change that the participants described as: "it feels uncomfortable", "it feels pretentious" "it feels as if I am a ballerina or wearing heels". Once we increased the lift of the same step, the experience changed to: "I feel like a horse... No, a flamingo".

Then I invite the group to walk backwards and to pay attention to the pathway of their feet. They realised that here the natural order tends to be from toe to heel. I suggest that they experiment with this walk and share their impressions. They discover that the bending is reduced and makes the walk difficult and very slow. The weight is distributed differently and creates a sense of danger as the direction of the walk is changed; the speed decreased as they could not grasp clearly all the movement around them. Supremacy of direct sight reduced, their peripheral vision became enhanced, instantly heightening kinaesthetic awareness and the relationship to the surrounding space.

I decide to break the order further by asking the group to keep walking backwards but to reverse the sequence of the foot, to move from heel to toe. As they try, their response is visible in their tensed facial expressions and in the decrease of their speed. They share with me that this experience is the most difficult so far as it requires lots of mental effort and concentration to achieve it. I ask them why? V. answers: "Because the functional order is broken simultaneously on two levels; you asked us to walk backwards first and then we changed the usual foot movement".

A slight or bigger change in movement transformed their inner landscape as well as the visible outcome. I could have asked them to 'walk like a horse' at the outset, however this intention would have activated, as it previously did when they copied R's arm swing, the external stimuli and the territory of 'I am moving or trying to move like', 'I am copying what I know', instead of intentionally creating the change using corporeal awareness, supported by the understanding of how movement works and how the *Instrumental Body* can affect and enrich my expression'.

Like ingredients in a kitchen, body parts and surfaces became ingredients within my *Instrumental Body* cooking arena. Awareness expands choices and therefore empowers the creator by patiently revealing more opportunities, hence more freedom. Tools from movement analysis supply additional ingredients to play with, ready to be mixed and matched for unique creative recipes. *The Instrumental Body* territory invites a playful mode and an experimental attitude.

"Play leads to brain plasticity, adaptability and creativity, nothing fires up the brain like play", says Stuart Brown (2008), founder of the National Institute for Play. He has studied the play histories of thousands of individuals and has concluded that play has the power to significantly improve education, personal health, relationships and the ability to innovate. When we play, we engage in the expression of our individuality, we expand our minds in ways that allow us to explore, to generate new ideas or to review old ones from a new perspective. Play is fundamental because it broadens the range of available options by challenging assumptions and giving permission to try untested territories. Play stimulates the parts of the brain involved in logical reasoning and unbound, carefree exploration.

Game 3: Breaking the relationship pattern

We begin a movement improvisation. From my movement kitchen, in addition to body parts, I give them more ingredients to play with, such as directions, levels and qualities. I also give them more freedom; they are allowed to keep their home body preferences, to play with them at times, to include other people's movement choices, or to break the order as much as their newly acquired awareness allows them. After a while I ask them to find a partner, then change and find another one. As they face each other, immediately all the idiosyncratic details suddenly transform themselves into a mirroring duet. We stop and discuss this change. All the pairs realise that each time they were facing each other, they were often smiling, adapting to each other's speed and keeping a similar 'not too close, not too far' distance. After a discussion about reasons triggering these similar choices, we unpack self-consciousness, habits, expected behaviour, motivation to please and to be accepted, cultural politeness, respect for personal space and boundaries. These are human needs expressed in our daily lives which are integrated in our Home Body.

This time I ask them to engage with their body as a relationship instrument and to deliberately play with distances, orientation, gaze, directions or levels. Progressively the relationship tableau becomes layered with going away, coming towards, facing away, side by side facing forwards, power games marked by changes of levels, indifferences indicated by non-meeting eyes, hopeful encounters suggested by long spatial distances accompanied by direct and durational gaze.

I bring in recorded lively music, containing no lyrics. I ask them to follow the musical rhythms or to go against them, to control their movement phrase endings by intentionally accelerating or decelerating. They tend to overuse their arms, so I remind them to use body surfaces and different parts. In partnerships, I suggest they use the space in between, around, above and below. Progressively the touch arrives, a foot gently touching someone else's knee, a forehead giving weight into someone's back. I suggest to them that they listen, include new choices, surprise themselves and the one they are dancing with. We have now transformed our pedestrian flavour into a stylised recipe. At the end of the improvisation G. says: "It was fun, I felt like a dancer, I lost track of time". *L. adds:* "my self-consciousness transformed into self-awareness".

We also relate to ourselves, objects, space and to other people by following a specific movement order. This obviously depends on cultural, emotional, health and other backgrounds but each of us uses specific elements of non-verbal communication when we do so. We are expected to face the person/people we are talking to in some way, occasionally nodding, expressing ourselves through a flow of gestures, postures and facial expressions, changes of distances and orientation between self and other(s). Some gestures like 'thumbs up' have a positive and encouraging meaning in some countries, but that same gesture is an insulting act in another for example. A head nod upwards and downwards means 'yes' in some places, but a 'no' in others. Once again, we use our body language naturally, as an embodied cultural code, and this allows us to interpret body expression in order to establish a productive communication with each other.

Body as an intentional instrument

To end this first session, I leave space for questions and I ask them to share their progression or findings. R. says: "My mind has been opened to so many possibilities", *V. adds:* "A body is really like an instrument, you need to play it every day in order to get better", *L. continues:* "I feel so grounded and so present. There is no end, I feel as I can go as far as my imagination will allow me to go", *M. shares:* "It was fascinating to experience how the expression changed in style during the session, from naturalistic, to physical theatre, animal studies to dance, just because we broke the functional order of movement".
V. asks: "What is next?"

In creative processes, devising workshops or classes I train participants to practise what I call the awareness ON/OFF button, moving from *Home Body* to *Instrumental Body* in order not to, automatically, take all their daily preferences, unconscious limitations, functional habits and patterns with them into creative work.

As *Home Body* and *Instrumental Body* are contained in the same body, although used for different purposes, this intentional awareness ON/OFF button tends to support the actor to detach from potential moments of doubt, self-consciousness, self-criticism, stress, anxiety, resistance or fear that could emerge during this subtle shift from daily life concerns into the creative process.

In this practice, awareness stands as a border checkpoint between the *Home* and *Instrumental* bodies. Instead of controlling a passport, it demands an internal consent to alchemise self-consciousness into a playful practice with the law of harmony. Practice or repetition enables difficult things to become easy. As we repeatedly do a task, the nerve cells make new connections. With repetition the connections strengthen, and it becomes easier for the brain to activate them.

By expanding sensorial awareness, by understanding the law of harmony and its movement organisation, by intentionally playing with movement ingredients and deliberately choosing to direct the body into new forms, dynamics and intervals, we develop new chemical stimuli and pathways for the brain, just like building new motorways.

Practising my awareness, by witnessing my patterns, by growing my 'movement kitchen' and by deepening my understanding and observation of the law of harmony, offers me tools for using my *Body as an Instrument* for different purposes. Depending on my work brief or context, I use the law of harmony differently. When my intention is to create harmony within the body systems, to promote health, wellbeing and mindfulness, I follow and emphasise the law of harmony and approach the body as an instrument for *self-regulation* to improve the body-mind connection by anchoring our living architectures into 'the here and now'. My awareness turns into a tiny door to the experience of a different time, a getaway into the state of 'being in the present'[3]. Expanding personal tendencies by playing with a toolbox offered by movement analysis could become a limitless game, an opportunity to grow our physical intelligence, a daily practice for growing sensorial awareness, a door to experiencing a life in *Kairos*. By practising sensorial awareness and working on our physical intelligence, we might develop an active listening with self and others, which could, in turn, inform our emotional intelligence. Or we might experience an enhanced relatedness to the environment, which might transform our relationship to climate change. Or we might just, playfully, enjoy our embodied selves.

However, when my intention is, as on this occasion, to use the body as an instrument for *performance, improvisation and devising work*, shining awareness onto the *Home Body* territory enables an intentional shift into *Instrumental Body* land which stretches body-mind relationship into playful territories of imagination and stylistic genres, enabling us to move, to think and to act more flexibly, creatively and unconventionally. A subtle or bigger degree of breaking of the law of harmony is permitted, as well as exploring all kinds of tools from the 'movement toolbox'.

In this case, my intention was to transform the director's idea into a visible form, to create idiosyncratic movement material, quick naturalistic transformations and stylised movement that could become the basis of an entire piece. I chose to use sensorial awareness as my fundamental tool to teach the company to understand how movement works and how our *Home Body* has an exceptional ability to become an *Instrumental Body* for stylistic investigations. We are now ready to explore, to mix or to juxtapose the personal *Home Body* patterns with idiosyncratic changes for the naturalistic scenes. As for stylistic needs,

---

[3] Chronos and Kairos were two words used for time by the ancient Greeks . The first concept is chronological time, the one that ticks, the one we measure. The latter is experienced in the now, only when we are fully present in the moment. One is quantitative, the other qualitative.

we will break the organicity of movement to create desired atmospheres and details. As Laban writes, "without a natural order within the single sequence, movement becomes unreal and dream-like" (1966: 6). Preston-Dunlop adds "inorganic can often create a surprise, an unreality" (2002: 67). A surprise in its turn generates a particular engagement with self or the spectator.

My intention being clear, my awareness glasses on, I know how to approach the law of harmony and what ingredients from my movement kitchen could help me cook a desired and, hopefully surprising, performance recipe.

Acknowledgements

Thank you to all the participants of my classes and workshops and to all my teachers, especially to Rosemary Brandt for hugely inspiring me and for supporting me in so many different ways, to become a choreologist.

References

Brown, S. (2008) *Play Is More Than Just Fun*, online at https://bit.ly/banda23

Brandt, R. (2007-2012) www.rosemarybrandt.com

Laban, R. (1966) *Choreutics*. Macdonald and Evans

Preston-Dunlop, V. (1979) *Dance is a Language, isn't it?* Laban Centre

_____ (1998) *Looking at Dances: A Choreological Perspective on Choreography*. The Noverre Press

Preston-Dunlop, V., Sanchez-Colberg, A. & Rublidge, S. (2002) *Dance and the Performative: A Choreological Perspective*. Verve

The Brain from Top to Bottom, online at http://thebrain.mcgill.ca

Olga Masleinnikova is an interdisciplinary maker, movement director, choreologist, creativity coach and lecturer. Her choreographic and collaborative works have been to venues in London, Ghent, Malmö, Dublin, Paris, Chur, Berlin, Reykjavik and Nanjing. Besides delivering numerous workshops internationally, she regularly collaborates with other makers and works on films. She has created an Artistic Professional Development Residency called 'Movement Toolbox' and coaches artists from all disciplines.

Her teaching portfolio includes Royal Academy of Music, Trinity Laban Conservatoire for Music and Dance, English National Opera for Opera Works, Middlesex University, Kingston University, East 15 Acting School, Oxford School of Drama, Bird College of Performing Arts, Cambridge University, JSBC Television China, HZT – University of the Arts of Berlin, Dance East, Le Grand Conservatoire d'Avignon and FOCAL Film Switzerland.

www.olgamasleinnikova.com

The Integrative Body

Awareness: the interplay of objective study and subjective experience

Elaine Hendry Westwick

Abstract

This chapter explores body-based awareness from two different perspectives: the objective study of the phenomenon of awareness and the subjective experience of awareness itself. Drawing on my scientific training and awareness practices, I explore what it means for subjective and objective ways of viewing awareness to co-exist. I write about what happens when the subjective experience of awareness, as refined through embodiment practice, permeates into the everyday act of living a life.

Introduction

This chapter explores body-based awareness from the objective and the subjective perspectives.

I define objective knowledge as that which is independent of an observer and verifiable by standards that do not vary between observers. The intention of scientific enquiry is to use all efforts to attain knowledge that is as close as possible to this ideal. What I call 'the objective study of awareness' places attention on the mechanisms and sensory components of awareness.

Subjective knowledge depends on something only accessible to the observer and as such cannot be verified by independent standards. Developing and deepening this type of knowledge forms part of the body-based awareness practices described here. What I call 'subjective awareness' places attention on the individual's in-the-moment experience.

I am a movement practitioner with a background in science. I draw on these experiences to explore subjective and objective ways of seeing awareness. I also explore what happens when subjective awareness is developed and distilled by awareness practices and feeds into everyday life. 'Life practice' is my term for the *doing of life* that is coloured by embodied awareness.

The subjective experiences of awareness discussed here are accessible to all but are honed by the formalised practice of an embodied art. The example used throughout is the practice of Authentic Movement, but there are many others, mostly falling under the umbrella of 'somatic practices'.

I learn Authentic Movement with Linda Hartley, who teaches Janet Adler's Discipline of Authentic Movement. In the practice, a mover follows the impulses of her body with eyes closed, often in the presence of others doing the same, all under the gaze of one or more witnesses. During the movement, mover and witness pay attention to their experiences, which may include images and feelings, and then afterwards come together to explore these experiences through language.

Authentic movement circle, Norfolk

> *I stand in the circle of movers and witnesses, my eyelids are already falling when the bell signals time to surrender to the impulse of movement. My awareness is focused inside, waiting.*

Almost immediately, a mover to my right starts jumping, at first tentatively and then with increasing delight. My internal witness is both alert to my unfurling body sensations and to the noisy joy nearby. I feel the soreness of my right hip joint, a familiar burn of unease. There is more weight falling on my left foot that my right. This is surprising, maybe it is how I stand these days. I hear laughter and feel the floor vibrate each time the mover lands.

I stay with the noise and pain and feel the wetness of tears on my cheeks. My nose blocks. I remain standing tall, motionless and open to the world, letting the tears be my movement. My internal witness is transfixed. My legs burn with the urge to jump and dance and giggle but I have been told to save my hips I must avoid "impact and load". I'm 39 and I can't jump.

The moving circle is a short one, perhaps 15 minutes, and the grief rolls for the whole time. I find myself rocking from my right leg to my left and back, small, slow movements, flooded with sensation. The movements feel both comforting and provocative. I track the rich internal experience as my weight shifts side to side. Noticing what is.

As the sound of the final bell fades I slowly open my eyes. Through the drying tears I meet the constant gaze of my external witness. I have been seen.

The objective study of the phenomenon of awareness

Awareness is the conscious experience of ourselves and our world in the present moment. It can be considered and studied from a third-person, objective standpoint and from a first-person, subjective one. I will start with an objective frame of reference.

The workings of the bodily senses are aspects of awareness that are amenable to objective measurement and study. Western science has traditionally privileged five senses: sight, hearing, taste, smell and touch. These senses gather information from outside the body using sense organs on the skin boundaries. The vestibular system, used to sense gravity, is often neglected and should arguably be added to this list.

Of more recent interest to science are the more elusive interoceptive senses, the senses of the internal body. The interoceptive senses include the inner landscape of sensation (for example intestinal contraction, heartbeat, pain, blood moving towards or away from a body part), the proprioceptive sense (the internal sense of position and space) and the kinaesthetic sense (the sense of bodily movement).

The external and interoceptive senses are blended by the brain to create an experiential gestalt, an awareness that is coloured by experience, memories and context. Much integrated sensory information remains unconscious and out of awareness; we are presented with what we need to know to survive in the world.

Processed data from the senses provide feedback from the internal and external environments which allow the body to respond, both consciously through decision making and unconsciously through reflex, in order to maintain a stable yet responsive internal environment. This selectivity of conscious awareness is itself partly conscious (we can choose to pay attention to the breath) or unconscious (we have no choice but to be drawn towards a fast-moving object).

In addition to studying the scientific foundations of awareness, the objective stance can also illuminate the underpinnings of awareness practices. The following components of awareness are explored objectively: attentional focus, attitude, relational awareness and repetition.

If awareness is like a patch of light illuminating experience, our attentional focus determines the location and size of the bright patch. To keep our bodies safe, we have evolved a focus of attention that is flighty and fluid. In survival terms it pays to be easily distractible if the distraction could be a threat to life. However, attentional focus is also adaptable and trainable and, through practice, a repertoire of awareness, which includes the stable, the narrow and the wide, can emerge.

The attitude brought to awareness practice is key to the nature of the awareness cultivated. To avoid slipping into rumination and distraction, an attitude of compassionate, non-judgemental curiosity is encouraged. It is accepted that wandering is what attention does, and the focus of awareness is again and again gently brought back to the present without becoming drawn into a mental story or planning the content of a post-practice lunch.

Awareness practice must also be placed in the context of interpersonal relationship. Practices that do not make space for the relational aspect of body awareness are more difficult to integrate into life. Authentic Movement explores relational body awareness through dialogue, ritual, witnessing, companionship and community. Embodied awareness arises in the presence of safety and ease; a sense of danger or threat is likely to lead to reactive hypervigilance instead.

Repetition is necessary to develop any skill, and the nature of awareness will grow to match the demands placed on it. Regular formal practice is necessary to keep embodied awareness spilling over into everyday life.

Autobiographical interlude

My formative worldview was one enlivened by the Western scientific method. In 1989 my dad gave me a copy of Richard Dawkins's The Selfish Gene, *which I read with near religious fervour during a glorious school summer holiday. An Oxford degree in Biochemistry and a doctorate in Protein Crystallography followed. The beauty of scientific exploration set my heart on fire, a love that still lives in me now.*

In my early 30s the arrival of a baby boy overturned my world and my science career slipped away. Now, many years later, along with a love of the rational, I am left with an almost pathological awareness of what I do not know, a feeling of profound physical discomfort in the presence of exaggerated scientific claims. Science is wonderful at disproving hypotheses, it is much trickier to validate a theory, particularly theories with strong personal meaning that involve complex, living, human beings.

By my late 30s I had an established business as a Pilates teacher, I had been slowed down by Feldenkrais and Alexander Technique, and yoga and qigong were important personal practices. It was then, in the depths of Norfolk, that I had my first taste of the discipline of Authentic Movement.

During the practice of moving and witnessing I felt self-conscious and out of my comfort zone in a rather compelling way. Here was an edge to explore, an intellectual challenge. Amidst the discomfort, there were also fleeting moments of presence and beauty. I was to return for more.

During the following years, my scientific core and my somatic practice bled into one another and my worldviews started to shift. Just as my rational outlook was part of how I lived my life, my somatic awareness practice began to change who I was.

The subjective experience of awareness

Turning to the subjective domain, awareness is a felt experience of 'what is', here and now. It is the ever-present hum of the body; the reason I know I am alive. Awareness is the experience of experience itself. It cannot be pointed to or measured, but it is real. The best I can do is to sense it, struggle to find the words to speak of it and engage with others as they try to do the same.

Awareness is not 'thinking about'. As I type, I bring my awareness to the pads of my fingers pressing the keyboard – not thinking about my fingers, but sensing their warmth, the hardness of the keys, their hesitant movement, the inner tingle that tells me they are mine. I have to think in order to communicate my experience, and there is the thought process that brings awareness to my fingers in the first place, but neither is the experience itself. While present with awareness, I don't feel I can choose whether further thoughts come up – they may or may not. Although, if I am paying attention then I have the choice to follow the thoughts or return to awareness.

In the moments when thinking melts away, the resulting awareness feels to me as simple 'being in aliveness'. I don't feel this sense directly in my fingers themselves as they type, I feel awareness of my fingers in the present. But in coming into the present moment I'm aware of a spread of pleasant warmth in my belly, which, if I can stay with it, may flow into something else. In a different moment, my aliveness may be one of dull, creeping heaviness and the accompanied sense of leaden resistance. Or it may be many things at once. But it is always alive.

Whilst the thinking brain is drawn to the past or future, the body forever remains in the present. The beauty of somatic practices such as Authentic Movement is that awareness is cultivated and carried as the body moves. The flow of awareness can take flight and expression in the current of bodily movement, with each moment of movement bringing new waves of sensory awareness to feed the flow.

As well as moving with awareness, Authentic Movement hones the skill of embodied witnessing. As I witness someone move with awareness, I pay careful attention to both their moving body and to the resulting experiences in myself. If I can sink into the moment and let go of 'thinking about' then I see this other as one who, like me, has awareness. I cannot know if the contents of their awareness are as I imagine them, although there will be the chance to explore this when we enter the conscious speaking part of the practice. But I sense that their experience of awareness itself may be broadly similar to mine – it is most likely that they feel the firmness of the floor, the light on their face, the movement of their chest as they breathe, just as I do. As the movement finishes, we look into each other's eyes, connecting through gaze what we seem to share so profoundly.

Along with exploring awareness in the relationship between mover and witness, the practice of Authentic Movement creates relationship between those who move, as the following vignette illustrates.

Authentic movement circle, Norfolk

I roll on the floor with my body long and open. I feel contrasting sensations of the fleshier and bonier parts of my body as they make contact with the floor. With my eyes shut I feel a sense of disorientation along with childish freedom to expand into the space around me.

My thigh rolls into contact with a warm firmness – another person. I pause, alertness flooding my body. Awareness draws outwards, from my interior to the sensations of contact with a responsive, living other. I do not know which part of which person I have contacted or her position relative to me, but my awareness now includes her. Thoughts jump up: Who? Where? What next? If I can allow the questions to rest back then there may be the opportunity to enter into a space of shared awareness. A space where I am present with my unfolding sensory experience, present with the touch, sound, smell and movements of her, and present with her awareness of me.

At the end of our movement together, I rest on my back with my pelvis on her belly and my knees bent in towards me. The weight of my lower body fully drops onto hers and she too seems to rest. In that moment, with my senses as sharp as my body is heavy, I am overtaken by what feels like a wave of awareness – here is my body, here is our contact, here is her body and here is everything else, right now.

Later, I speak with my witness and try to open more and more into that moment of connected awareness – our precise body positions, the mover's belly moving as she breathes, the back of my head on the floor. I search for the place in my body where I felt the sensory experience of the wave, hunting for where it began. Something jars, the place seems elusive. Then I know that the wave is all of my body and all of everything else and it comes from outside, although it is still me. On speaking these words I feel the rightness of knowing this is as close as I'll come to describing the experience for now.

The meeting of the subjective and the objective

Both objective study and subjective experience are valid perspectives on awareness, each true as far as they go but incomplete without the other. Awareness itself is non-conceptual and exists regardless of whether it is split into subject and object.

Despite their differences, there are many overlaps between science practice and movement practice. Both derive from the same impulses – to creatively experiment with the world, to chase the unknown, to clearly observe, to repeat and modify until clarity or further questions emerge, and to explore and refine in relationship with others. Both are driven by the pleasures of curiosity and passion, of knowing and understanding and of being part of a community with shared values and outlooks. Both part of my experience of the world.

Despite efforts to the contrary, the subjective domain inevitably forms part of science practice, from the decision of what to study to the motivations, personal and cultural history and values of the practitioners. The reverse is also true – the objective domain is found within somatic practices such as Authentic Movement: there must be a structured form of practice, the cultivation of a compassionate rather than vigilant awareness and a careful use of words to try to capture and communicate personal experience.

I do not believe that my experience of awareness outside my body means that there is a physical ether waiting to be discovered. I sense that the experience is derived fully and only from my body in the presence of other sentient bodies. For me it is marvellous and awesome that my body can create a sense of awareness which spreads beyond its own boundaries, and even more so that others may share this experience too.

At home, Cambridge

July turns into August, and I have five full days of slowness at home. The children are with my parents, I move the garden lounger into the living room and make a "not to do" list. To be able to walk slowly around the house, without anyone needing me and nothing to do is startling.

In the kitchen I become aware of how habitual urgency marks my movements. A breakfast bar juts into the room and, even without the children pulling me back and forwards around it, I swing round its corners.

In the space of solitude I notice this familiar movement for the first time, one I already know deeply in my hips. In order to circle the breakfast bar efficiently I walk a tight circle, sliding my pelvis out to the side, away from the counter top. If the pelvis moves sideways and the legs stay upright then then one hip joint tilts in and the other out. I slow the movement down, analysing it, feeling it in my body. When my right hip is tilted in it hurts. I realise my body has known pain in this movement for years but this is the first time it has been named.

The discovery takes a few days to process and, over that time, I find myself playing with alternative ways around the barrier – ones where my feet move more and my hips less. Over these days, miraculously, my hip starts to hurt less when I'm sitting on the lounger. The discovery is delightful – with space to bring my awareness to what my body already knows, there are possibilities of healing. This is a corner that I can no longer cut.

Life practice

Gently and resolutely, often without conscious intention, awareness practice permeates into everyday life. The infusion is hazy and indistinct, to help convey its essence I differentiate between awareness volume and awareness location.

Awareness volume indicates how much attention is taken up by sensations from inside and outside the body, compared with attention drawn by cognitive processes of thinking, analysing, planning and remembering. In an Authentic Movement practice the volume of awareness is turned up and dominates over thinking, but this is not a practical nor desirable state as I interact with my children, plan the week ahead or teach a Pilates class.

Location refers to the physical positioning of the awareness. It may be somewhere precise or indistinct within my body, or I may become attuned to the awareness of another person or to the wider world. It may be in more than one place at once. To some extent I can choose where to place awareness – as I type now, I feel the pull into awareness of the screen and my thoughts of words to come, but I also choose to have a low volume awareness of my shins on the floor, the cushion under my thighs and the pleasure of my breath. When talking with another person my awareness will automatically read the social situation, but I may be able to expand my focus to include the presence of my body and their sense of being.

Awareness is not always pleasant: it is opening to the unknown of 'what is'. When my son wails about a perceived injustice, I may sense in myself an internal turning away with the words 'it doesn't matter' coming to mind. It's not always possible, but if I am able to accept his emotions as real and mattering deeply, I see those emotions are given the chance to move and transform. When another person is able to hold his discomfort, my child can begin to find his own inner compassionate witness.

Refined awareness brings to daily life the option of choice. By being aware of the interplay between external events, bodily responses, cognitive appraisals and emotionality, it may be possible to regulate cognition and behaviour to promote better health. That is, to move from an automatic way of being to one of responsive, active engagement with the world.

There may be concerns that enhancing body awareness can lead to obsession with bodily functions, creating somatosensory amplification and anxiety. This is true if the awareness is a type of protective hypervigilance responding to threat. But if the awareness is present moment, curious and compassionate then it is not thinking or ruminating or worrying and it can edge towards an acceptance of what is. We may then find that there is wisdom in our sensations, that they contain truths which may or may not be welcome.

Questions and Surprises

Any exploration of the crossovers of the subjective and the objective inevitably generates more questions than answers. Must we practise ourselves in order to study a movement practice? How then to deal with the inevitable bias? Can the scientific method ever evolve to study the richness of subjective experience? Is it possible for the subjective to become more valued in science? Could the objective add more rigour and accountability to somatic practices?

In addition, any exploration of awareness in the context of life practice inevitably provokes unexpected personal realisations. I finish with three from my experience.

First, as I'm more aware of my sensitivity and intense relationship with life, I am also more aware of inconvenient discomfort that indicates when I need to stop and change activity.

Second, I am surprised by a sense of the sacred or numinous that is seeping into my rational, sceptical bones. There is no feeling of conflict, for the sense of mystery comes from within, it is immanent in my body and a part of physical reality, rather than supernatural and other-worldly.

Finally, I notice that what I formerly saw as objects I treat more as subjects like myself, affecting how I interact with and care for them. Increasingly, I see inherent value and worth in everything that makes up our planet, I feel the pain of treating the environment as an object, and, fleetingly, I have the odd sense of being a part of something bigger.

Further Reading

Adler, J. (2002) *Offering from the Conscious Body: The Discipline of Authentic Movement.* Simon and Schuster

Batson, G. and Wilson, M. (2014) *Body and Mind in Motion: Dance and Neuroscience in Conversation.* Intellect Books

Hartley, L. (2004) *Somatic Psychology: Body, Mind and Meaning.* Whurr

_____ (2015) 'Choice, surrender and transitions in Authentic Movement: reflections on personal and teaching practice', *Journal of Dance & Somatic Practices*, 7(2), 299-312

Hodges, P.W. and Tucker, K. (2011) 'Moving differently in pain: a new theory to explain the adaptation to pain'. *Pain*, 152(3), S90-S98

Mehling, W.E., Wrubel, J., Daubenmier, J.J., Price, C.J., Kerr, C.E., Silow, T., Gopisetty, V. and Stewart, A.L. (2011) 'Body Awareness: a phenomenological inquiry into the common ground of mind-body therapies'. *Philosophy, ethics, and humanities in medicine*, 6(1), 6

Mallorquí-Bagué, N., Garfinkel, S.N., Engels, M., Eccles, J.A., Pailhez, G., Bulbena, A. and Critchley, H.D. (2014) 'Neuroimaging and psychophysiological investigation of the link between anxiety, enhanced affective reactivity and interoception in people with joint hypermobility'. *Frontiers in Psychology*, 5, 1162

Weiss, S., Sack, M., Henningsen, P. and Pollatos, O. (2014) 'On the interaction of self-regulation, interoception and pain perception'. *Psychopathology*, 47(6), 377-382

Yakeley, J., Hale, R., Johnston, J., Kirtchuk, G. and Shoenberg, P. (2014) 'Psychiatry, subjectivity and emotion – deepening the medical model.' *The Psychiatric Bulletin*, 38(3), 97-101

Elaine Hendry Westwick is an independent practitioner-researcher-educator based in Cambridge, UK. Her career began in biomedical research (PhD in Biochemistry) and in 2005 she changed focus to study movement, embodiment and self-awareness. Elaine is a certified Pilates and Qigong instructor and studies Authentic Movement with Linda Hartley. She teaches community movement classes and interdisciplinary workshops and is a Registered Somatic Movement Educator with ISMETA.

www.embodiedscience.com

The Intuitive Body

Feeling my way: walking in darkness, fog and whiteout

Margaret Kerr

Abstract

In our culture, lighting is ubiquitous at night, and it is not normal to make a journey on foot in the dark, in thick fog or in a whiteout. Although navigating in low visibility is familiar to outdoor leaders, the sensory experience of this practice, and its emotional and spiritual dimensions, are rarely spoken about. In this chapter, I will draw on my own experiences to explore these dimensions, and to show how they can teach us about our relationship to each other, to our culture, to the earth and to ourselves.

We are slogging up a tussocky hill with partial snow cover in freezing conditions, heading for an impenetrable cloud. No obvious reason to be here from my perspective. Lots of stops while my companion takes bearings and looks at the map. Then we enter the cloud. Snow flurries. Nothing to see but grey at 10 metres.

And so it goes, as we struggle on upward with no view and nothing recognisable. Counting as we go. Interminable sequences of 65 steps (75 if steep). This is a visual junkie's hell, and no prospect for the top. Remarkably, we seem to hit way marks with regular precision. Just as well, as a glance behind shows no trace of our progress, as the wind obliterates our steps.

The most worrying is the complete change of direction manoeuvre required at particular map-confirmed points. What if that was the wrong rocky outcrop? Will it still be recognisable on the way back down? I forget what sequence of steps I am counting. Was it 5 x 65 or 6?

At last the trig point, the summit and big smiles. Don't get me wrong, I am elated. But mostly relieved. Absolutely nothing to see. The high point is an ice encrusted iron fence with spicules of ice on the lee of the post, like long sharp finger nails.

And so back down. My other high point is recognising the rocky outcrop, having counted correctly this time.

(G, first experience in a whiteout)

Not being able to see the way ahead in a wild landscape can be frightening and disorientating. In our times, it is not normal to make a journey on foot in the dark, in thick fog or in a whiteout. Lighting is ubiquitous at night, and our culture is designed for those who can rely on their eyesight. Some jobs, for example, farming, wildlife conservation or police work may involve walking over fields or hills at night or in the snow, but this is a necessity, not a recreational choice.

Learning to navigate in poor visibility is part of the training for outdoor leaders in the UK. Thick fog, low cloud or snow are often part of outdoor adventures, and a leader must be able to guide their group to safety whatever the weather. A long route, or unplanned events can mean having to find your way home, or to a safe resting place, in the dark.

Although navigating in low visibility is familiar to outdoor leaders, the sensory experience of this practice, and its emotional and spiritual dimensions are rarely spoken about[1] Here I draw on my own experiences to explore these dimensions, and to show how they can teach us about our relationship to each other, to our culture, to the earth and to ourselves.

Around 600m, the snow firms up underfoot. At about the same altitude, I go into the mist and it starts to snow.

White. Only white. No shapes. No colours. Points of light dance all around me. My body tells me I am going uphill. My breathing becomes faster, my heartbeat quickens. My thigh muscles feel more load. The tendons at the backs of my feet are stretched.

I can see my feet and legs through the baseplate of my compass. There is nothing else to see.

I walk on a bearing of 320 degrees and count my paces. My breathing eases where the contours should widen.

Everything is white.

Everything is...

I realise how peaceful I feel. My body is so quiet inside. No grasping. No tension. I am part of everything.

My feet are level and my breathing is steady. This must be the top.

The ground starts to take me downhill. I adjust the bearing. Turn right. North along the gentle ridge. I know there will be steep slopes loaded with snow on the left and cliffs on the right but my compass shows me the way I can trust.

Slowly, the glen starts to emerge from the whiteness. Mountain hares with grey-white coats are bounding around in the snow. The hillside below is brown and speckled with white. A y-shaped track full of snow winds down the hill, leading me home.

---

[1] Blind explorers and travellers such as Amar Latif, Erik Weihenmayer and Andy Holzer have done much to promote outdoor recreation among people who are not able to see. However, sadly, as yet, there is little literature on the phenomenology of outdoor experience among this population.

This is a description of an experience I had walking through a whiteout – where snow in the air and on the ground had obscured all the landmarks. Navigating at night or in dense fog has a similar feel, although here, vague shapes of the surrounding land are often visible. Whiteout puts the visual sense into the background by flooding the eyes with light. Night does something similar by immersing the eyes in darkness.

Once the visual sense is stilled, other sensations come to the fore. I am brought to feel the shape and texture of the land directly through my body. The materiality of the place I am in, my own materiality, and the connection of my body to the earth become palpable and vibrant.

In whiteout conditions, I rely on sensations from my body to tell me if I am going up or down hill, as I cannot see the slope of the ground. I can relate this sense of slope to the spacing of contours on the map. The spacing and rhythm of the contours act like a score for my heart rate, my breathing and my steps – frequent contours mean a steep slope, short steps, frequent breaths and heartbeats: widely spaced contours mean flat ground, long steps, widely spaced breaths and heartbeats. The shape of the land is transduced into the rhythms of my body.

Walking in whiteout and in darkness, I have vividly experienced shifts in the location of my awareness – the conscious sense of self which both directly witnesses, and *is* experience. This witnessing self feels fluid: at the same time, a process and an entity. It concentrates in various parts of the body as I move (for example, head, heart, lower abdomen or limbs), or extends to the environment around.

In the whiteout experience I described above, my awareness was dispersed, and I felt this as being 'part of everything'. It is as if the extreme whiteness around me drew my attention out and released me from grasping at any one part of the experience. This brought a sensation of peace, openness and quietness to my body. I felt held in something greater than my self, and safe to relax into that luminous space.

However, it was the training in navigation and my trust in the compass bearing that made this possible. The compass becomes an extension of my body. I hold it parallel to the ground, pointing forward in front of my heart, bent elbows braced at my sides. For accurate navigation, the compass and my body must move as one.

The compass lets me follow a magnetic line. Like the thread that Ariadne gave to Theseus to guide him out of the labyrinth, it brings me safely home. If I had not had this guidance and this training, my experience would have been very different. I would have been lost and alone, in danger of falling over a cliff or straying onto avalanche-prone slopes. My awareness would have gathered into a fearful ball in my chest, or scattered fragmented outside of me, fretfully searching for clues.

No matter how well prepared or confident I am in my skills, there are times outdoors at night, in snow or fog when I feel afraid. At these times, attending closely to where the fear comes from, watching how it arises, spreads and disperses through me helps me to know what to do. Sometimes the fear comes in response to an anxious thought about weather or snow conditions. Sometimes, it comes from a child part of myself that is afraid of being alone in the dark. Sometimes it seems to come from the place I am in: areas of land that feel haunted by war or grief. Turning towards the fear and finding out where it is coming from can help me to take the right action – to turn back or change my route in response to the weather, to comfort the wee one in me who is afraid, or to acknowledge that maybe some suffering has happened in this landscape.

A full discussion of the experience of feeling suffering held in the land is beyond the scope of this chapter. However, my experience is that it is easier to attend to the subtleties of this kind of experience when my mind and the land are quiet at night. Sometimes, like the shaman's drum, the rhythm of my walking opens up a deeper layer of awareness. It is as if the skin between the past and the present becomes porous. And my own skin becomes porous. I start noticing currents of emotion that I do not recognise – coming from the land itself; from the air around me, or up through the ground. A miasma from layers of history soaked into a place, but still volatile enough to feel.

When these feelings come I sometimes stand still, quietening the movement of my body as if to hear something that is beyond the normal human register. Sometimes, the strength of these feelings demands a decision – do I stay in this place and let them come more fully into me? Or do I move away and continue on the route I have planned? Sometimes I stay and switch on my head torch. Like awakening from a dream, the light goes some way to disperse the intensity of what I feel.

Generally, I have noticed that if I switch on my head torch, the light seems to draw my awareness out from my body, along the beam, and into the pool of light cast on the ground. If I walk in the near darkness, awareness rests in my body. If I walk with a head torch switched on, it goes ahead of me, on my future route. The torch beam pulls me forward, unbalances me a bit, and I can feel slightly separate – a few steps ahead of myself.

Of course, there are times when walking with a head torch is really helpful – where there is no ambient light from the moon, the stars or a local town. In these very dark conditions, if I leave my torch turned off, the air feels thicker on my skin and in my lungs. Sudden darkness, for example on entering a forest, feels like a wall in front of me, pushing my awareness back into my chest, like a shock wave. At other times, when I rest, still, in the intense blackness, it feels velvety, soft, and womb-like. At these times, I am aware of the darkness inside my own body, quiet, receptive and warm.

My experience of awareness in the near-dark of night walks has a different character from that in the light of a whiteout. I feel more grounded and bodily in the dark: more earthy and immanent. Sensations from my ankles and legs pull me into the dark interior of my body when I am walking uphill. On uneven terrain, the feeling is more in the soles of my feet, close to the texture of the ground. Walking steadily on the flat, my paces are even and rhythmical. I feel my mind still and calm – resting in the base of my abdomen – what I know from the martial arts as the *hara*. If I am walking downhill in the darkness, the awareness rises up and I can feel a sensation of lightness in my chest, a feeling of nearly flying.

The weather adds an extra layer of bodily experience in the darkness. A strong wind makes it hard to rely on my habitual sense of balance. What was automatic becomes effortful, as I have to keep adjusting to the push and pull of the wind. Sometimes the wind overwhelms my hearing with its relentless noise, and I long for it to stop: when the constant buffeting dies down, it can leave a profound sense of peace in its wake. At other times, when the wind stops, I can feel strangely wistful, as if I have lost a companion.

Extreme cold renders my mind scattered and muscles sluggish, or draws the focus of awareness single pointedly to painfully cold hands or feet. The weather can radically alter the feeling tone of being outdoors in low visibility. Velvety darkness turns spiky with the addition of wind and rain. Soft snowy whiteness transforms into a maelstrom of spindrift: tiny stinging shards of ice pelting the body from every direction.

If I am on my own, my awareness is focused keenly on the sensations in my body, and those I feel coming from the land around. If I am with companions, I am also focused on them. The experience of making a journey alongside others at night, in fog or in snow, can bring a deeper sense of intimacy and trust. The lack of visual distractions in the wider environment often strengthens the bonds of feeling within a group or pair, and shared vulnerability can help companions drop their defences and open up to each other. However, in extreme environmental conditions, fear and uncertainty can also amplify pre-existing conflicts and stresses between people. Navigating these difficulties takes skill, patience and delicacy with each other. Our bodies can help us with this. For example, with G's first experience in a whiteout, I was aware that I needed to keep a sense of my own centre, a quiet trust in the process that I felt somewhere in my lower abdomen. At the same time, my heart felt open and tender to how much he was struggling. Listening to both these places helped us stay together and make the journey safely.

In near-darkness or fog, many of the normal features of our industrial and agricultural age are erased or rendered uncanny. Sometimes, it is hard to tell from the land what century I am in. Although in many places, farming and forestry have changed the landscape radically since the Neolithic period, in others, I am only aware of the shape of the land, unchanged since the ice retreated 11,000 years ago.

Sometimes, I am suddenly and unexpectedly brought up against the marks of our culture. Pylons loom up unexpectedly out of the fog, festooned with barbed wire and warning notices; the electrical fields from power lines pull on the compass needle and distort direction; farm animals appear as eerie shadow beings; city lights irradiate the sky, glowing ethereal, or post-apocalyptic.

When these familiar elements of our 21st-century world are transmuted by darkness, I have a chance to glimpse things differently. Electricity is no longer just something that comes out of a wall socket and powers the TV: it is an elemental force. Cows are no longer sources of milk and meat, they are mysterious, quiet, steaming animals – relatives of the wild aurochs that roamed Europe in prehistoric times. When I look at the glow from the city I can see how we are burning up fuel throughout the night.

Reduced visibility shifts my sense of proportion as I travel through the landscape. A hillock which is 10 metres high can take on the proportions of a small mountain when I see it filtered through fog or snow. Often it is only when I start to climb the feature that I realise its true size. I can feel suddenly too small, too large or too clumsy. If I start off with any tacit illusions of mastery over the terrain or the journey, these soon have to give way to a careful moment-to-moment process of feeling my way.

Journeys in these conditions make me aware of my vulnerability, but they also they offer a way of being that helps me take care of myself. When I see less, I feel the flow of sensation from my body more acutely, and this becomes my guide. I can feel my way even though I cannot see the route ahead.

This experience of feeling my way, guided by bodily sensations is a useful touchstone at other times in my life where there is uncertainty, or I cannot see the way ahead. Remembering how I can rely on the compass to get me safely home also reminds me that it is ok to let go of ideas about where I think I ought to be going. When I rely on the earth's magnetic field, I put my faith in something trustworthy, invisible and much greater than myself. My body is vital in both situations. It registers the subtle currents of a greater flow, whether in relationships with others or with the land. It helps me understand more about what is there, even if it is unseen. It helps me balance the different aspects of a situation and shows me the way to go.

The outdoor experiences I have described can bring a meditative quality of awareness. Especially in whiteout, bodily experience becomes more vivid and the focus needed to navigate quietens my mind. This brings about a quality of awareness that I recognise from Buddhist practice – something like *Samatha*, a steadying and calming of the mind, and *Vipassana*, a vivid insight into the nature of phenomena.

I want to explore in more detail this meditative aspect of walking in reduced visibility, taking as my guide the philosopher Shigenori Nagatomo's (1992) theory of attunement between body, mind and environment.

Rather than seeing intellectual and bodily knowledge as irreconcilably separate, Nagatomo describes how 'knowing-that' (intellectual knowledge) and 'feeling-that' (bodily knowledge) can be gradually integrated through somatic practices such as sitting meditation or karate. The somatic practices which Nagatomo describes as bringing about this integration form a *'kata'* or disciplined pattern of action. The *kata* minimises external sensory stimulation and calms the mind. This lets the body's subtle language come into the foreground. Learning to navigate in snow, fog or darkness could be seen as an example of learning a *kata*.

Until the *kata* becomes familiar, there is a tension between the mind and body: the discipline feels awkward and negative thoughts loom large. However, with repeated practice and familiarity, mental and physical turbulence settles. This allows implicit, felt bodily knowledge to gradually become explicit, and to reach a harmonious balance with intellectual knowledge.

Importantly, Nagatomo's theory does not just confine this harmony to the skin-bound self. The integrated awareness that he describes pervades the 'living ambience' of a person's environment and all the objects therein. Most importantly, he characterises this awareness as 'bilateral', suggesting that it is reciprocated by our surroundings. This is the condition of 'felt inter-resonance', where we can sense the depth of our environment through the depth of our bodies, and our environment, in its own mysterious way, resonates with the depth of us.

In the description that opened this chapter, G had never navigated in low visibility before. It was his first time practising the *kata* of being in a whiteout. There is an understandable tension between his mind, his companion and the environment, and his bodily knowledge is muted.

In contrast, I am more familiar with this *kata*, and the description of my experience in a whiteout shows more integration of bodily knowledge, intellectual knowledge and environment. My mind has quietened with the limited sensory stimulation:

> *I realise how peaceful I feel. My body is so quiet inside. No grasping. No tension.*

And in the feeling that '*I am part of everything*' there is a sense that the everyday dualism between myself and the place has dissolved.

It is interesting to apply Nagatomo's theory to my experiences of feeling suffering in the land at night. As sensory input is quietened by the darkness, and my body falls into the familiar *kata* of counting steps and following a compass bearing, the conditions become favourable for a felt inter-resonance between myself and my surroundings. As in other meditative states, the normal sense of time can start to fall away. Through the bilateral attunement between myself and the place I am in, the land may then begin to disclose the secrets that it holds within itself.

Nagatomo suggests that if we can open to the depths of our body and let body knowledge come together harmoniously with intellectual knowledge, we can start to release and heal any suffering which is held within us. This is congruent with the work of understanding and releasing bodily trauma in psychotherapy. However, following Nagatomo's theory, it is clear that this act of coming together is not confined to our own interior. There is a bilateral attunement as well – between us and our environment. If this is so, then perhaps opening also to a felt inter-resonance between ourselves and our surroundings will help to release and heal any suffering which is held in the land. If we can become comfortable in the dark, then we can turn towards that suffering, and feel our way towards what we might do to help.

> *As I walk across the moorland in the fading light, I begin to experience a sensation of pressure in my chest. It mounts to my throat. I realise it is intense grief. It builds and builds until the pressure feels intolerable.*

Then I start thinking of the women who starved to death along the shore here during the Highland Clearances. With that thought, I start sobbing. I kneel down and press my forehead into the sodden ground.

By now it is dark. My hands find the stones in my pockets that I gathered from the shore earlier. The stones are cold and I know I need to warm them up. Now I know they represent the women, and maybe even more than that, hold something of their spirits.

The grief subsides and gives way to warmth in my chest. I stand up and switch my head torch on; make my way back to my tent near the shore. There I light a fire against the cold, arrange the stones close to the fire and sit with them until we are all warm.

I make porridge and offer some of the oats to the fire.

My body feels peaceful. The stones are warm to my hand. Something has settled, perhaps even healed.

Acknowledgements

I would like to thank fellow explorers Joe McManners, Rebecca Crowther, Roland Playle, Simone Kenyon, Karen McMillan, Daisy Martinez, Sue Mcleod and Patrick Earle and the instructors at Glenmore Lodge, particularly Carl Haberl and Al Gilmour for teaching me to navigate in the dark. I would also like to thank Natasha Lushetich for her academic input and for introducing me to the work of Shigenori Nagatomo.

Reference

Nagatomo, S. (1992) *Attunement Through the Body*. SUNY Press

---

Margaret Kerr works in Scotland as an artist, psychotherapist and outdoor educator. Her work is focused on strengthening connections between people and the rest of nature, and on weaving together different ways of understanding the world.

@megkerr245

The Moving Body

Sandra Reeve

Abstract

This chapter investigates movement art practice through the lens of embodied/metaphoric cognition, exploring how the awareness that arises through movement can challenge fixed concepts, habitual attitudes and personal blind spots. As well as this, movement practice is described as one possible way of rekindling a relationship with the right hemisphere of our brains, seen as a necessary step in these heartless times of global fragmentation, competition, dehumanisation and collapse.

Sometimes movement is flat – it has no nuance, no sensing, no impulse and our receiving of the world and of the other remains flat. How can we wake up our recognition in the pool of life, wake up our understanding, our awareness and our sensorimotor life so that we can feel 'not flat'?

Awareness itself is not flat. It can be likened to a piece of fabric that is not taut. If it is taut it cannot breathe. It loses its 'living' nature. It loses sensitivity, and that is 'flat'.

(Suryodarmo, 2014: 311)

Introduction

I start by describing the lineage of *Move into Life* because it provides the context for my emphasis on *movement* and the *moving* body within the broader arena of body and awareness. Despite its transience, movement creates a field in which bodies and environments reveal their mutual interdependence: the shift of my body's weight on the earth, a gust of wind, an outbreath, a gesture through space, the flap of a wing...

Move into Life is a movement practice which I have developed over the course of my life. My primary influences include Jerzy Grotowski, *Satipatthāna* and *Joged Amerta*.

First of these was my early work with Jerzy Grotowski in Poland in the 1970s which prioritised an attitude to physical training as a process of unlearning. It also referred to the actor-in-flow as a channel for the transformation of perception. The later paratheatrical work and 'theatre of sources' sought to shift perceptual realities by direct involvement in (ritual) actions, both traditional and contemporary.

Secondly my training in *Satipatthāna*[1] with John Garrie Rōshi fundamentally influenced my experience of the moving body through the practice of the Four Foundations of Mindfulness and an understanding of *Paticca-samuppāda*[2]. This gave me a direct experience of our habitual mechanisms, attitudes and conditionings as revealed *through our bodies and through our actions*. These practices shifted my awareness towards *how* I was doing things, rather than *what* I was doing, and to the experience of change as the only constant factor in life.

---

[1] *Satipatthāna*: a way of mindfulness.

[2] *Paticca-samuppāda*: the Law of Dependent Origination which deconstructs the mechanism of habit formation into twelve separate stages.

Finally, since 1988, I have been influenced and inspired by *Joged Amerta* with Suprapto Suryodarmo (also known as *Amerta Movement*[3]). *Amerta* is a Javanese word which he translated as the 'nectar' or 'elixir' of life. This practice starts from the basic movements of daily life: walking, sitting, standing, crawling and lying down and the transitions between them, beginning with the observation of children playing. It is also based on moving in nature and an embodied study of movement from the play of elements in motion and the expression of life in nature. "In Javanese traditional thinking there is no sharp division between organic and inorganic matter, for everything is sustained by the same invisible power" (Benedict Anderson cited in Foley, 1995: 163). This cultural attitude is manifest in *Joged Amerta*: the changing environment and all beings-in-movement are utterly interdependent and the mover is invited to move freely and to develop an awareness of the influence of her own position and of her own 'com-positions' within the flux of life.

So, my practice can be seen to be influenced by the potentially transformative relationship between movement and perception/attitude; by the reality of change as the only constant; and by movement as a profound link between humans, their environment and other sentient beings. *Move into Life* offers a basic training in movement. The idea of a training in movement often puzzles people, so let me elaborate: movement is both a skill and an art form. Movement as a skill can be applied or referenced in many areas of life: for example, the performing arts, somatic practices, therapy, sport, bodywork, or within daily life. Movement is also an art form, with its own beauty and (like dance) has its own aesthetic value, expressing an idea or emotion or responding to an event.

Early morning and I stand noticing the room around me. I pay attention to the sounds, the smell, the colour, the texture, the shapes, the atmosphere, the entrances and exits, the windows. I turn around slowly, becoming aware that I am well rested after a good night's sleep. I look up at the vaulted roof and seeing the space above me, I follow the impulse to stretch and then yawn as my body bends slowly

---

[3] In more recent years, Suryodarmo called his own practice 'Joged Amerta' to include 'moving-dancing' within the name of his practice. (*Joged* is a style of folk dance in Bali and a common term for dance in Indonesia.) This new name also created a soft distinction from the work of many students and colleagues within a growing community of practitioners who began to adopt his original term 'Amerta Movement' to describe the lineage of their own work with him.

and stretches the muscles on either side of my spine, and the back of my calf and thigh muscles. I feel the stiffness in my ageing limbs easing and find pleasure in sinking to the ground and onto my back where I continue to stretch , release and rest in the pause between those two motions, without losing the sense of my surroundings.

'Receiving my condition' is the first stage of becoming aware of my body-in-movement-in-the-environment. As I move, I begin to notice the physical sensations within me and to be stimulated by the details of my surroundings. I may also become aware of my other senses, of thoughts or of a predominant feeling – if I do then I try to incorporate those into my movement so that they have expression rather than remaining internalised and invisible. I move at the shifting threshold of paying attention to my moving, sensing/perceiving body and paying attention to the structure, the actuality of my context – in this case to include the roof, the floor, the windows and doors.

This quality of taking time to receive the condition of my body, the environment and the atmosphere around me, including other people, before 'doing' anything or following an impulse is a distinctive aspect of *Move into Life*. It allows me to pay attention to a wider experience of where I am. It may be seen as the equivalent of listening before needing to speak, of feeling for the 'volume' of a situation in all its complexity through noticing details and from that position allowing action to emerge or taking action.

However, receiving and expressing occur together through movement. The body moving as it becomes aware of its condition within the changing environment is already expressing itself. It is participating in what is happening, rather than standing outside as an observer, and thus I am gradually becoming aware of the impact of my own presence on the situation. As part of my approach to movement I make 'change' rather than 'stop' the default, so I experience stopping or pausing as occurring along the line of movement, rather than at the 'beginning' or 'end' of movement. Movement includes stillness. I experience the movement(s) within me and around me as ever-changing and never ending.

As ways of enlivening my body-in-movement-in-the-environment, I give myself little tasks – the 'opening' and 'closing' of the joints, creating different shapes in space, finding variations in shaping the body's constellation of limbs at different levels in the space by lying down, crawling, crouching, sitting, standing/walking/running and jumping. Sometimes I play with the gradation of movements, from the smallest movement of a finger to a full-bodied lunge through the space. I begin to notice my habitual patternings of movement, familiar sequences ... and I notice the different, the fresh and the new. Gradually the little tasks fade away and I am just moving: following the impulses that arise within or that are stimulated by the world without and noticing what I notice within and without.

Here I engage with my joints and muscles as a way of connecting directly with my sensorimotor system, playing with many variations to refresh that system by moving differently and to widen my movement vocabulary. A question which has been my long-standing friend, in a performance context, in a therapeutic context and in daily life is: 'how can I engage with my own habits in movement so as to be able to embrace a wider set of choices?'. In daily life, the way I move is constantly conditioned by the environment that I am in and my body's movements are constantly conditioned by familiar, habitual ways of doing things. In my practice I can explore different combinations of flow, weight, time and space as I move. At the same time (or, instead) I can move to find a deeper acceptance of restrictions that exist in my body by fully inhabiting what I *can* do and by 'placing' or 'putting' my movement in the space, just as I might play a chord on the piano or make a mark with a paint brush.

> As a way of approaching being embodied, I like the idea of our movement itself as a costume or as our clothing. Clothing includes the sense of beauty, design choice, filtering and an individual's signature in the signs of nature. We are all just part of an environment.
>
> (Suryodarmo 2014: 310).

Movement appears to be fluid and intangible – it is very different from an experience of 'body' – and yet if I take the time to feel the movement as it is happening, I can become aware of the impact of that movement both on my organism and on my surroundings. At the very least there will be a change of shaping and of atmosphere and this I can register as an influence on my subsequent moves.

I know my movement patterns well; I recognise the tendencies, qualities and sequences that are in motion and that can repeat themselves, apparently often regardless of the changing environment (despite everything that I have just said about mutual influence). I experience those particular movements as the ones that stay hidden from awareness the longest, that I seem to be most identified with, that I am most familiar with and that I call 'me' or 'mine'. They have had, and may still have, an important role to play in my life, they can be deeply reassuring as well as limiting my choices.

Now if we look through the lens of the cognitive sciences: "Contemporary neuroscience seems to suggest that concepts [...] make direct use of the sensory-motor circuitry of the brain" (Gallese and Lakoff, 2005:19). So perhaps it makes sense that the patterns which repeat themselves most through my movement will also connect with my 'habitual concepts', the concepts that I am most attached to, both personally and culturally. For example, if I value 'dialogue' as a concept I will move very differently both with others and in the environment than if I value either 'harmony' or 'challenge' as a concept. (Of course, this is a very simplified example – in practice, I don't just value one thing or move in one way.)

Moving towards the more recent field of *embodied* cognition and the view that "human cognition is grounded in sensorimotor processes and in our body's morphology and internal states" (Ionescu & Vasc, 2014: 275) – I can see that my preferred patterns and sequences of movements speak to me of my own preferred ways of perceiving the world and acting within it, some of them learnt when I was a child.

Next, following the thread of of Claus Springborg's research into embodied cognition and learning processes, if I begin to merge the view of embodied cognition with cognitive metaphor theory, I could say that how I move, my unique 'clothing' in movement, must also embody my 'cognitive metaphors' (Springborg, 2018: 81). By this I mean that the language I use to refer to my world will also be reflected in the way that I apprehend the world through my body in motion.

If a metaphor is an expression that describes a person or object by referring to something that has similar characteristics, then "cognitive metaphors are the systematic use of experience from one domain to understand and engage with another domain" (ibid: 72) and they always highlight the viewing lens and ignore other possible lenses.

So, for example, if I use metaphors from cookery to describe the modules of my annual Project Group, cookery becomes the 'source domain' for me to understand the structure of my course. It will also influence how I speak and think about the course, *how I act within the course* and it will prevent me from viewing the course in other ways. Were I to use a metaphor of 'conquering outer space' to describe the phases of creating an individual movement-based project within that Project Group, it would be an utterly different experience for all concerned. So linguistically, the image I use will affect my attitude and my actions. But we can take this one step further if we realise or remember:

> ...that the first domain of experience is our sensorimotor experience of being a body in a three-dimensional space... that we naturally have experiential dimensions, such as front side vs. backside, up vs. down, and center vs. periphery, and that these fundamental experiential dimensions are what we ultimately use to ground our understanding of all other domains.
>
> (Lakoff and Johnson in Springborg, 2018: 81)

This means that our embodied experience of environment is the ultimate reference for our understanding of everything else. This, in turn, implies that by exploring our movement vocabulary with appreciation, curiosity and *with a contextual awareness of place and time*, we are stimulating cognitive flexibility in our apprehension of the dimensions that we use to ground our understanding in other fields, rather than fixating within a limited pattern of experience.

Our movement, the language we use, the metaphors we choose and our memories have all been created by our relationships to our respective environments and to each other and are all uniquely woven together and present in our living, moving bodies.

As I move, paying attention to my structure, sensory input and the shaping of my body through space, I am stimulating my metaphoric and associative life and I am noticing what presents itself to me from my internal archive of memories and habitual concepts. I am once again in a condition of receiving before initiating (or passive before active)[4], where the action emerges from first taking the time to become aware of myself in the situation. These associations may appear as thoughts, words or images which I embody or I may feel the impulse to make certain gestures or sounds, to speak or to sing as I move, perceiving and then evoking, for example, some past atmosphere, landscape or experience.

> *I notice the brightly coloured bunting, hanging from the golden beams;*
> *I notice the spaces between each little triangular flag, and the way they*

---

[4] One of the four movement dynamics (Reeve, 2015).

*waft in the breeze coming through the window. I notice the geometric motifs, the floral motifs and how the **inscribed** multicoloured leaves match the shape of the ficus plant's dark green and fleshily **incorporated** leaves, standing in the corner of the room with a home-grown avocado plant. Since 1978, I have always had a home-grown avocado plant, wherever I have made myself a home. Quite simply, it has come to represent 'I am at home', and I experience a surge of warm feeling as I acknowledge that meaning, as I squat beneath its leaves for a moment. I take the time to examine its leaves with a fresh eye, I smell the leaf, I stroke my face with a branch of leaves, I breathe and relax for a moment as past memories arise from this detailed sensory contact and then I follow my movement into the centre of the room.*

I am paying attention to the particularities of my actual surroundings which may present themselves to me in different layers: as what they are in the concrete world (i.e. bits of triangular material hanging in the room); as 'bunting' (the name that those objects have been given in my culture); as a set of prompts and associations (they were part of the wedding celebrations in a village hall in Dorset and, going further back, they remind me of Tibetan prayer flags, evoking memories of my travels in Ladakh and of spiritual devotion).

By staying for a moment, by touching and imitating the quality and texture of leaf through my body's subtle movements, by imaginatively feeling into the leaf and noticing how it attaches to the rest of the plant, how it lives in a patterned proximity with the other leaves, I have begun a layered relationship with it. This all takes place in the present moment, concurrent with associations arising from the past, which have been stimulated by our contact in this context. I am perceiving the actual plant itself but it also has a metaphorical life for me. So now my moving body is stimulated both by aspects of the present environment, by past memories and by the spontaneous meanings that I attribute to the gestalt of this moment of movement. Movement can incorporate many layers of experience at the same time. It is the joy of a pre-verbal or non-verbal medium like movement that it can hold all of these realities, all of these descriptions and that they can comfortably co-exist; in fact, their co-existence is 'the fact', and our attempts at reductive linear descriptions may be a poor substitute for the appreciation of the complex 'whole'.

If we don't pay attention to the whole, we may begin to believe our own partial descriptions and to mistake them for the whole 'truth' because of the value our society gives to our brain's left hemisphere's capacity to focus on details and parts, to look narrowly for information, to 'unpack' what it experiences and to come up with 'explanations' for it.

Iain McGilchrist, psychiatrist and writer, in his book *The Master and His Emissary*, investigates in detail the strengths and weaknesses of each hemisphere and tells us how they need to work together. He argues that the left hemisphere has come to predominate, contributing to the evolution of the contemporary social values that drive us, and overriding the less tangible attributes of the right hemisphere. I shall return to this imbalance later, and to his observation that it matters not just *what* we attend to in any given situation but *how* we attend to it. (McGilchrist, 2018)

Continuing for now with the field of embodied cognition, how we attend to something is conditioned by what Springborg has called our 'sensory templates' (2018: 57). By merging the ideas of embodied cognition and cognitive metaphor theory discussed earlier, we can see that "we use our sensorimotor experiences as templates upon which we build our understanding of abstract phenomena" (ibid: 93).

I'll use my left hemisphere to unpack that! A 'sensory template' is the sequence of actions, embedded in my sensorimotor system, that I use to represent a concept to myself. For example, 'I am buried in paperwork, 'I am fighting popular opinion' or 'I am drowning in information' all carry with them a particular flavour of how I am experiencing my world. Someone else might skip through their paperwork, surf the wave of popular opinion, or devour the information. If I am asked to say how I know *bodily* that 'I am drowning in information', I might say, for example, that I feel as if I cannot breathe, I feel a sense of pressure above my head and as if I am both stuck in one place and being carried away at the same time. Now I have some actions that I can move with, first to acknowledge and to accept them and then to explore possible alternatives (an active approach) or I might recognise these qualities or tones of movement when they arise in my daily practice/life (a passive/receptive approach). Either way, change now seems possible and I can have a less fixed sense of self. By recognising these 'sensory templates', I can begin to see how I have a position from which I can open them up or let go of them – or from which I can disturb my implicit biases[5] by engaging with a different set of actions/movements in relation to a particular concept or attitude.

---

[5] 'Implicit bias' (also called 'unconscious bias') refers to attitudes and beliefs that occur outside of our conscious awareness and control.

Our implicit and typically unconscious conceptions of ourselves and the values that we live by are perhaps most strongly reflected in the little things we do over and over, that is, in the casual rituals that have emerged spontaneously in our daily lives.

(Lakoff and Johnson, 2003: 235)

At the same time, I can notice the relationship between the word I use to denote a concept, the concept itself and its sensory template. For example, the words 'to go' will evoke a concept and then a set of actions and feelings in me: *to stride out, open space ahead, no looking back.* (I may not perform these actions every time I hear the words, of course, but I experience their reality in the way that a pilot experiences the exhilaration of flying even in a simulator). The words 'to leave' will evoke another concept and set of actions and feelings in me: *antennae active, to turn away, a feeling of stickiness behind me, no looking back.* The words 'to be left' stimulate yet another concept and set of actions and feelings in me: *crumpling, freeze, deaf and cannot see.*

All three sets of words (and sets of actions and feelings) offer me information about myself, about how *I* respond to those words and the concepts they represent. The nuances of my response may shift according to where I am at the time, for example at an airport, a friend's house or a cemetery but the underlying, embodied experience of each concept, remains immediate and constant, wherever I am. Those visceral experiences mark the fact that different sensory templates form the basis of my emotional life experience of those variations of transition. It is only by recognising, accepting and changing the underlying set of movements that condition how I experience those dynamics that I can change my experience of them and therefore their emotional content. This is the crucial point: following acceptance, changing the language alone remains unconvincing; but if I can also change the sensory template, then transformation through a new set of choices is possible.

> Standing in the centre of the room I continue to move, noticing that I have become aware of the sounds around me and that my movement has softened as it registers the sound of the sea. I am standing beneath the point where the two ribbons of bunting cross over. I feel myself to be at a crossroads and from there I can see in many directions by turning 360 degrees: a golden yellow wall with a wood burning stove; an expanse of sea meeting the horizon in a blend

of greyblue; many, many books and some old Chinese blue and white plates balanced on a high shelf, with, improbably, a little square window in the top left hand corner of the wall, which no one can look through; a sideboard full of photos of my mother, my parents, my brother, my grandparents, my great grandparents, all interspersed with odd ornaments and buddha figures from around the world.

Suddenly, as I move, I enter more fully into the associative realm, just lightly aware of the room now. I feel myself to be at the end of a long line of ancestors. I recognise their gestures, their facial expressions and their stances in the photos. I feel proud and bemused – they are deeply known and total strangers and I am the last in this line of beings. I am moving rather than sitting still or meditating; my movement is reading the situation as it presents itself to me / as I choose what I am attending to; my body-in-movement is involved in both reading and expressing. And I am also a 'witness' to the passing expression of my body-in-movement because I am connected to my sense of the space within me and to the space around me as my movement takes shape through space. I remember the story of the Buddha who left the palace and his family to seek enlightenment and liberation from the Wheel of Samsara, longing to see the world as it really was. I wonder, briefly, if these two paths, the Path of the Ancestors and the Path of the Buddha can in fact co-exist or if they are very different ways of being in the world?[6] Do I need to let go of my attachment to my family to reach enlightenment or…? And then I notice that my body, apparently of its own accord, has gone down to the ground, that I am kneeling and bowing towards the East, facing the wide expanse of open sea.

Here, at one level, my experience of being at 'the crossroads' opens up a sequence of associations around 'staying' and 'going'[7] which, when I was moving, I linked with life choices and with the process of ageing. In my associations of ancestors and Buddhist practice, I seem to have returned to the topic of 'making less the habitual' so that I can be freer in my movement, liberated from conditionings that keep me from being present within the constant changes in the world around me. My movement piece finishes facing the sea and I have a sense of peacefulness and belonging, with all the ancestral photos at my back.

---

[6] The Buddha left his family to seek enlightenment along the Path of non-attachment.
[7] Another movement dynamic (Reeve, 2015).

Stop. Hold your position for a moment. Allow yourself to soften and to relax inside. Breathe. Relax inside.

At this point, I would like to return to the question of what we attend to in our lives and perhaps more importantly, *how* we attend to it (McGilchrist, 2018) as crucial questions for the moving body. We have seen that playing with variations in movement with an awareness of the changing environment opens up new movement and cognitive possibilities. This approach to movement practice also offers new ways of *attending* to our experience, prioritising the right hemisphere to balance the "triumph of the left hemisphere" (ibid).

McGilchrist maintains that attention "is actually nothing less than the way in which we relate to the world" and uses the fable of the Master and his Emissary to describe the relationship between the left and right hemispheres of the brain. In a nutshell, the right hemisphere is the Master – it sees the whole before the parts. It alone can see the broader picture. The left hemisphere is the Emissary who gradually usurps his Master, believing that he knows everything through the power of fragmentation, logic and abstraction whereas in an ideal world he should be unpacking things in this way only to hand it back to the Master for reintegration into the experiential whole.

Today, as a result of being taken over solely by the ways in which the left hemisphere experiences the world, we find ourselves in a world which has become, according to McGilchrist "a heap of bits": increased specialisation, information replacing knowledge, an increase in abstraction and reification, a lack of gradation and proportion as living becomes more mechanical and more focused on material things; a depersonalisation of relationships, reasonableness replaced by rationality and a "downgrading of non-verbal, non-explicit communication". The left hemisphere, operating on its own, lacks a comprehension of intangible qualities such as common sense, social cohesion and the bonds that can exist between people and between people and places. [8]

We can see from this description that movement as a practice can support the nature of the right hemisphere by attending to a world where there is 'betweenness': a non-verbal, pre-reflective world. Movement encourages us to be in that world as we receive it, move with it and express

---

[8] My synopsis of the left and right hemispheres draws heavily on Iain McGilchrist's elegant choice of words over several pages in his short book *Ways of Attending* and the whole synopsis here should be seen as a 'direct' quote. My intention is to relate his observations to movement awareness.

ourselves within it. A practice of movement may be seen as a profound act of recovery of many of the values attributed to the right hemisphere which include: an appreciation of time as flow, rather than as a succession of moments; an experience of a visual depth of field; a rekindling of empathy and of most aspects of our emotional life (with the exception only of anger which is more associated with the left hemisphere). Re-establishing a relationship with the right hemisphere of our brains by giving value to the intelligence of the moving body may be seen as a necessary step in these heartless times of global fragmentation, competition, dehumanisation and collapse. Remembering that the simple act of paying attention to the moving body holds the potential for transforming deeply seated biases and attitudes may be important when words are not enough.

Conclusion

> Movement is intrinsic to human expression. It precedes and underpins cognition, language and creative art. Movement may be seen as the most fundamental 'skill' and one that adapts to the environment.... If movement is the 'skill' of skills, through which we act in the world, and are acted upon, then a study of movement should reveal the process of my unique becoming-being-becoming in the world.   (Reeve, 2008: 237)

In this chapter I have explored movement through the lenses of embodied cognition, cognitive metaphors and 'sensory templates'. If all my concepts are grounded in the sensorimotor system then it is evident that, as I move, these ways of speaking and thinking about the world (which are also the ways that I am in the world) will come into consciousness, and as I move they will re-present themselves in some way, either through the quality of the movements or by evoking images, metaphors or associations. I can get to know my selves and my preferred 'lenses' by becoming aware of how I move in different situations. An awareness of context is as important as the awareness of my actual movement. I can also choose to accept my movement and/or to initiate other patterns of moving in the world.

At the same time, because I am in the flow of the unknown when I move and because I am cultivating an awareness of the broader picture (that is to say I am attending to the world through the right hemisphere of the brain by moving with an awareness of the sensorimotor system), I can also cultivate a space of stillness and presence within movement as witness to this everlasting transient arising of embodied associations and memories that I often believe to be me, relaxing any fixed or determined sense of self.

References

Bloom, K., Galanter, M. and Reeve, S. (eds.) (2014) *Embodied Lives: Reflections on the Influence of Suprapto Suryodarmo and Amerta Movement.* Triarchy Press

Foley, K. (1995) 'My Bodies: The Performer in West Java', in P. Zarrilli (ed.) *Acting (Re)Considered: Theories and Practices.* Routledge

Gallese, V. & Lakoff, G. (2005) 'The Brain's Concepts: The role of the Sensory-motor system in conceptual knowledge', *Cognitive Neuropsychology,* 22(3-4)

Ingold, T. (2000) *The Perception of the Environment.* Routledge

Lakoff, G. & Johnson, M. (2003) *Metaphors We Live By.* Univ. of Chicago Press

McGilchrist, I (2018) *Ways of Attending.* Routledge

_____ (2019) *The Master and his Emissary.* Yale University Press

Nyanaponika Thera (1983) *The Heart of Buddhist Meditation.* Rider

Reeve, S. (2010) 'Reading, Gardening and the Non-Self', *Journal of Dance & Somatic Studies,* 2(2), 187-201

_____ (2015), '"Moving Beyond Inscription to Incorporation": The four dynamics of ecological movement in site-specific performance' in V. Hunter (ed.) *Moving Sites: Investigating Site Specific Dance Performance.* Routledge

Springborg, C. (2018) *Sensory Templates and Manager Cognition.* Palgrave Macmillan

Suryodarmo, S. (2014) 'Interview' in K. Bloom, M. Galanter, and S. Reeve (eds.) (2014) *Embodied Lives: Reflections on the Influence of Suprapto Suryodarmo and Amerta Movement.* Triarchy Press

_____ (2019) Joged Amerta Programme (pdf)

Ionescu, T. & Vasc D. (2014) 'Embodied Cognition: Challenges for psychology and education', *Procedia – Social and Behavioral Sciences,* 128, 275-280

Sandra Reeve is an Honorary Fellow at the University of Exeter, UK, where she completed her doctoral thesis on the Ecological Body in 2009. 'Move into Life' is her cyclical programme of autobiographical and environmental movement workshops in West Dorset. Her movement research is influenced by walking, complexity thinking, sustainability, meditation, gardening and performance. She both facilitates and creates small-scale ecological events, as well as mentoring individual movement-based creative projects. She is trained as a movement psychotherapist and supervisor.

www.moveintolife.com

The Pain Body

Awareness and embodied walking meditation: a self-reflexive approach

Jamila Rodrigues

Abstract

In this chapter I show how awareness plays a role in reshaping ideas of physical pain. By awareness, I mean the natural ability that each one of us has to imagine, visualise, and focus into the sensations, feelings and emotions that occur in the body and mind. In other words, awareness in this chapter is seen as the 'here and now' of experiencing walking meditation in the natural environment through a body in pain, as it happens. This chapter suggests that walking meditation may be approached as a somatic practice (instead of a spiritual practice) and that scholars can critically engage with events and topics that impact on their personal and professional worlds, while navigating between those spaces.

Feeling ill and falling down

> *In 2015 a series of 'strange' symptoms began to 'attack' my body affecting my career as a professional dancer. Certain muscles and organs of the body began to inflame without apparent reason. At this stage I was still able to dance, perform and live a normal life with the aid of medicine and alternative therapies. A year and half later the inflammation process exploded into a full body inflammation. I was bedbound for six months, unable to walk, eat or to perform hygiene routines without help. I was carried in a wheelchair to visit numerous specialists and tested for Multiple Sclerosis, Alzheimer's, Rheumatoid Arthritis and other conditions. Still with no diagnosis after 18 months, I tested positive for an undetected viral infection that affected my central and peripheral nervous system. I was left with visceral inflammation, pelvic fibrosis and cognitive dysfunction. But I still had a body.*

Vignette 1 – Buddhism, Bhikku Bodhi

> Again, bhikkhus, when walking, a bhikkhu understands: 'I am walking'; when standing, he understands: 'I am standing'; when sitting, he understands: 'I am sitting'; when lying down, he understands: 'I am lying down'; or he understands accordingly however his body is disposed...

> Again, bhikkhus, a bhikkhu is one who acts in full awareness when going forward and returning; who acts in full awareness when looking ahead and looking away; who acts in full awareness when flexing and extending his limbs; who acts in full awareness when wearing his robes and carrying his outer robe and bowl; who acts in full awareness when eating, drinking, consuming food, and tasting; who acts in full awareness when defecating and urinating; who acts in full awareness when walking, standing, sitting, falling asleep, waking up, talking, and keeping silent.   (Namoli, 1995)

Buddha's teachings suggest that awareness is vital to reach an enlightened state; in other words, no human action should be unaware if one is to sustain mindfulness through life. Meditation is generally regarded in Buddhism as a process of getting to know, and working with, the mind; allowing the mind to rest in its natural awareness in a state of non-distraction (Simpkins & Simpkins, 2013).

The body also plays a crucial role in other forms of meditation, such as walking meditation. Monk Jack Kornfield[1] explains walking meditation as "a universal practice for developing calm, connectedness, and embodied awareness".

For example, in Japanese Zen Buddhism[2] walking meditation (*kinhin*) is often practised clockwise around a room, with specific gestural postures and takes place between *zazen* (seated meditations). *Kinhin* meditation teaches that the body posture follows specific rules such as the left hand creating a fist with the thumb tucked inside, while the base of the thumb presses on the solar plexus and the right hand covers the left hand. The upper half of the body remains in the same posture of seated meditation, while the lower abdomen and neck are relaxed, chin tucked in, and the gaze lowered.

In Theravada Buddhism, monks practising walking meditation do not have a set of positions. Instead, they can walk for many hours, often outdoors, as a way of developing concentration and attention to the sensation of walking and letting go of thoughts.

Zen Master and peace activist Thich Nhat Hanh developed his method of walking meditation using affirmations such as 'I am solid' or 'I am free' in order to encourage a positive mental state while walking. He refers to walking meditation as a means to cultivate awareness, "the practice of walking meditation opens your eyes to the wonders and the suffering of the universe. If you are not aware of what is going on around you, where do you expect to encounter ultimate reality?" (Hanh, 2011: 64)

Despite variations, the overall goal of the praxis is to learn how to be aware when walking, using the natural body movement to encourage mindfulness and a wakeful state.

I took the structural form of walking meditation as a guideline and applied it to walking in nature as a platform for pain awareness and pain release. I chose to do this practice, which increases awareness of physical sensations, as part of an embodied self-reflexive process. While I focused attention on the body movement sensations, I also used visual stimuli such as colours, words or situations. By focusing on the 'here and now' of walking as it happened, the sensation of pain began to take a different shape. I paid attention not only to the body pain but also to how the feet move and the variety of sensations and perceptions of the present moment.

---

[1] https://jackkornfield.com/walking-meditation-2/
[2] http://www.meditation-zen.org/en/node/1335

Vignette 2 – First steps

To be able to walk and be autonomous after such a traumatic period of my life is a journey into acceptance and refusal, adjustment and re-adjustment, getting better and relapsing. Throughout this process my ideas of body pain shifted, resulting in a new awareness of walking mindfully.

I am standing again. What is left?

"Ways of moving are ways of thinking" (Sklar, 2001: 4)

The manner in which we sense ourselves knowing the world and our environment is, as dance ethnologist Deidre Sklar suggests, also a way of thinking. I call this 'interconnection' (an interconnection between the sensorial, cognitive, emotional and corporeal) a somatic process of being and acting in the world.

Living organisms are somas: that is, they are a necessary and ordered process of embodied elements which cannot be separated either from their evolved past or from their adaptive future. A soma is any individual embodiment of a process, which endures and adapts through time, and it remains a soma for as long as it lives. (Hanna, 1976: 31).

Practitioners in the field of somatic studies took Thomas Hanna's idea of 'individual embodiment of a process' and understood soma to be a union of mind, body and spirit. This idea is the key to the individual's self-knowledge. Somatic thinking gave birth to numerous therapeutic approaches. Martha Eddy, somatic pioneer, spoke of these approaches promoting a connection between 'human potential' and notions of 'holistic health' (Eddy, 2009: 5), whereas Glenna Batson, researcher into embodied cognition and Alexander Technique teacher, suggested that they "offered a whole new language of consciousness and body wisdom through self-awareness and self-guidance" (Batson, 2009: 2). Being engaged with and attending to this dialogue between the body and mind, people "can learn newly, become pain-free, move more easily, do our life work more efficiently, and perform with greater vitality and expressiveness" (Eddy, 2009: 5).

Within this context, walking meditation fits well as a somatic process, a method and platform from which people act, sense and relate to themselves, to their attitudes and to their bodies. For example, when instructing on walking meditation methods, Thich Nhat Hanh suggests:

> Walk upright, with calm, dignity, and joy, as though you were an emperor. Place your foot on the Earth the way an emperor places his seal on a royal decree. A decree can bring happiness or misery. Your steps can do the same.　　　(Hanh, 2011: 90)

These visual exercises are just one part of walking meditation. Other bodily exercises within walking meditation are, for example, the learning of conscious relaxation through breathing or the use of phrases and repetition to induce movement or self-reflexivity. These features are components used to assist the practitioner in the embodied experience. Similarly, somatic modes of attention enable people to understand 'how' the body in movement is known, instead of attending to "what world realities are factual and what is belief" (van Ede, 2010: 70). Following van Ede's idea, the fact of falling ill and the reality of the physical pain informed my body movement, as did my attention or awareness to the 'how'.

As I began to improve, I decided to begin a practice that involved both body movement and engagement with the natural environment. The method of walking meditation, as I saw it, was ideal because it not only gave me the confidence to be alone again but also helped me to shift pain perceptions, while retraining the body back into movement. The natural setting, as in nature walks, brought me a sense of calmness and gave me a solitude that was comfortable and safe for me.

To be able to write about this experience in a reflexive, critical manner, I began to ask myself: What sort of awareness does my body experience in walking meditation? How is pain to be pictured and felt if one is to develop awareness? What benefits might this practice bring to me personally and how might reflecting and writing about this practice afterwards be of interest to those working with somatics, body and pain?

Vignette 3 – Walking

> *If my feet move, I cannot feel it, as if an imaginary thread was held like a hook from the foot to the coccyx, passing through the vertebrae towards the skull ... as if it were glued, and I do not feel.* (anatomy)

> *As I walk I pay attention to this space of 'I do not feel', directing my attention to the point of contact that I once felt and knew as sole of the foot.* (felt sense – staying with) *And I cannot breathe. I imagine that sole of the foot I once felt and danced on many times, I imagine the lungs opening and closing, filling with oxygen, but I do not feel.* (respiratory system – visualisation/imagination).

Right now I stopped imagining and was taken by the green of the leaves, still hanging from the tree, waiting to fall. (merging experience)

Suddenly, in the place of a foot lives a little light waiting to be light out. What I've been before is no longer, now I am aware I was already whole. With or without foot. There are trees surrounding me. There is one in particular that stands before me. This tree had fallen down and grown back again. (reflections from environment/intimations of acceptance) *The connective tissue that keeps the memory of a past life, spontaneously remembers that it is not forgotten.* (anatomy/body memory) *And I walk on through the forest...*

As a former dance practitioner, I was able to build on my own 'somatic repertoire' – a mixture of previous dance exercises and somatic awareness, including breath, colour and stretching. For example, I focused on a particular part of the body such as the lungs as I breathed, becoming consciously aware of them. Sometimes I would choose a colour to connect to an organ: for example, I chose blue for the lungs, magenta for the uterus, and yellow for the liver, brown for the kidneys and so on.

Colouring lets us focus on a particular part of the body and 'continue colouring' throughout the body. On a certain day, wherever my connective tissue felt particularly inflamed, I would 'send out more colour' to the tissue, as a helpful 'anti-inflammatory' colour therapy. I drew imaginary maps of muscles working and of tissues freeing up, often imagining a spider's web that needed unfolding or stretching or a gentle touch.

With basic knowledge about the meridian system[3] I focused on the meridian route of a particular organ/body part that would stretch alongside that channel. Here I use Bob Cooley's[4] 'resistance through stretching' technique, which teaches us to contract and to resist the contraction while stretching generates more flexibility and strength while eliminating fascia tension. The 'meridian stretches' and 'meridian mashing' are a series of exercises to free up energy flow and help with scar tissues and adhesions throughout the body.

The outward elements also played a crucial role. I would listen to the sound of birds, gazing at the dogs walking beside me, sensing the sun or

---

[3] Meridians are regarded in Chinese medicine as channels of energy that run throughout each organ/body part.

[4] https://www.thegeniusofflexibility.com/resistance-stretching/

the rain and the wind. I would pay attention to the pathway and the levels of difficulty or ease while moving through the forest. Before stepping, I would assess the level of difficulty and make a conscious decision either to step in or back. Interestingly, my courage levels would increase if I watched the dogs walking in front of me, as if they would show me where it was safe to go. I also varied the times I walked. This might be early morning, other times at sundown, sometimes alone, sometimes with the three labradors.

I tried to pay attention to the physical sensations, the thoughts and emotions that arose from practice and not to focus on the actual pain as a negative consequence of my health situation. I noticed how my body shifted to adjust to the pain that was felt at the time of walking, yet, sometimes attending to the pain by touching a trigger point or going deep into the muscle tissue was necessary. During this time, I had to make conscious decisions: do I stop and attend to the pain? Do I attend to the pain but continue to walk? Do I ignore the pain and continue to walk? All of these options were experimented with and analysed.

For instance, when I attended to the pain and was 'forced' to stop, I would continuously and consciously choose to remain with an outer focus (on a tree, a bird, the sunlight). This conscious decision as an exercise was to keep awareness in balance between pain (inner) and the elements (outer), not allowing pain to take control of my wellbeing during the practice. When I kept walking while massaging certain trigger points, I noticed that pain patterns began to change as the body needed to adapt to the action (walking) and an inner stimulus (being massaged) at the same time.

With every somatic praxis I chose to do there are transformative changes within the body: breathing, awareness of bodily sensations and the body's reactions, touch, movement, chanting or gestures. All emerge from and within this complex relationship of addressing physical pain and a state of awareness which allows you to remain calm despite the pain.

When I chose to continue walking without attending to the pain, I could sense the levels of pain rising, as if in 'protest' at not being attended to. This last option was by far the most challenging. I noticed that the body and mind required a higher level of focus in order to press the 'ignore' button and carry on. When the levels of pain were eight or nine out of ten, I would begin to breathe deeper, walk slowly and move more carefully. Without ignoring the pain, I would set my mind to guide me, as if a third voice was talking to me, guiding the process. This moment is perhaps where I felt the most spiritual connection, as if the 'other' (Nature, God, Spirits, Ancestors) came to life to help me through the walking. The so-called 'third voice' can potentially be regarded as the point where full awareness performs at the highest level.

Vignette 4 – as it happens

> *As I walk, I begin to sense the world through heat, itchiness, discomfort, muscle tightness, change of heart rate, breathing. The effects of these sensations are captured internally as well as externally, from the environment. The (my) environment is the surroundings in which we (I) and other organisms live and develop. There's a tree that fell down and continued living. I stop here, and I look at the tree. I focus on the roots and where they came from, and I look up following the trace of the branches and leaves. Fall and stand-up. Not straight position, but in the somehow harmonious bend shape, as if the tree is bowing, at life.*

A new world of possibility arises when I/practitioners turn inward and reflect on the embodied experience itself in a non-judgemental way. This kind of considered reflection makes the experience more detailed and conscious and usually happens after the experience is completed. After the walking experience I gave myself self-reflection time and self-analysis and used my dance skills to ponder the different aspects of experiencing pain while engaging in walking meditation as a somatic practice. As a practice of reflexivity, I tried to recall the levels and patterns of pain and the body movement that emanated from it, the changes, the re-positioning and positioning of the body, adjusting and re-adjusting while trying to remain honest to my inner feelings and to sensations (Ellis & Bochner, 2000) that were sometimes quite overwhelming, depending on the level of pain.

This type of introspective approach is challenging, since the analysis and the writing happen after the experience. I found that the best way to describe an intimate moment that had happened while walking often left space for vulnerability, doubts, or even fear and emotional confusion.

By pondering on how the experience was, we can effect change. This change is associated with my ideas of pain. If I can perceive that experience differently, I can also change how I act or react to pain, as in a so-called negative, positive or neutral way. The impact for the future is a consequence of my reflection. My dance and somatic studies background tell me that the full benefit of this experience cannot be fully felt if one is not grounded in the body. In other words, as pain levels increase or reduce, the bodily sensation shifts from mere discomfort to awareness. While I note these changes occurring both in my mind and body, pain patterns change quality and intensity, simply through my being fully aware.

Once these ideas are conceptualised differently, I sense a receptiveness and openness to the corporeal sensations that emanate from the action of walking in unison with the whole surrounding without focusing on the actual physical pain. I called this process 'beyond the body pain' experience as a consequence of the practice.

Conclusion

The method presented here can potentially be helpful for those studying topics related to body, body movement, embodiment and somatic studies. The case study has shown, for example, how dancers who can no longer dance can learn creative ways of dealing with body pain and cultivate awareness while maintaining a steady relationship between body and mind.

My discussion of walking meditation as a somatic practice can be a resource for dealing in an alternative way with traumatic events and exercise awareness. The unpleasant experience of dealing with any kind of pain, and here I focus on physical pain, can lead the individual to operate in a disembodied state. Without focus, one can easily fall into a circular state of suffering, anxiety, sadness and frustration. In my case, this circle is no longer enclosed in body pain but forced me to open up and question other aspects of human life such as humility, patience, courage, empathy and, most of all, resilience. For a former dancer living in pain, it is a journey into cultivating these qualities and also into opening to a new kind of awareness that can be simply achieved if I consciously 'watch my toes' as I walk forward.

Acknowledgements

I am grateful to Sandra Reeve and Sara Stuart for their patient readings, criticism, suggestions and editing work. To 'Laura' my grandmother, her spirit never rests.

References

Batson, G. (2009) 'Somatic Studies and Dance'. International Association for Dance Medicine and Science. Available at: www.iasms.org

_____ (2011) 'Integrating Somatics and Science', *Journal of Dance & Somatic Practices*, 3(1-2), 183-93

Eddy, M. (2002) 'Somatic Practices and Dance: Global Influences', *Journal of Dance & Somatic Practices*, 34(2), 46-62

_____ (2009) 'A Brief History of Somatic Practices and Dance', *Journal of Dance & Somatic Practices*, 1(1), 5-27

Ellis, C. & Bochner, A. (2000) *Autoethnography, Personal Narrative, Reflexivity: Researcher as a Subject,* in N.K. Denxin & Y.S Lincoln (eds.) *Handbook of Qualitative Research,* Sage, pp. 733-768

Hanh, T.N. (2011) *The Long Road Turns to Joy: A Guide to Walking Meditation.* Parallax Press

Hanna, T. (1976) 'The Field of Somatics'. *Somatics Magazine – Journal of the Bodily Arts and Sciences,* 1, 30-34

_____ (1979) *To Dance is Human: A Theory of Non-Verbal Communication.* University of Texas Press

Namoli, B. (1995) *The Middle Length Discourses of the Buddha: A Translation of the Majjhima Nikaya.* Wisdom Publications

Reeve, S. (2011) *Nine Ways of Seeing a Body.* Triarchy Press

Sangharakshita (2004) *Living with Awareness: A Guide to the Satipatthana Sutta.* Windhorse Publications

Simpkins, C.A. & Simpkins, A.M. (2012). *Zen Meditation in Psychotherapy: Techniques for Clinical Practice.* John Wiley

Sklar, D. (2001) 'Toward Cross-Cultural Conversation on Bodily Knowledge', *Dance Research Journal,* 33(1), 91-92

Smears, E. (2009) 'Breaking Old Habits: Professional Development Through an Embodied Approach to Reflective Practice', *Journal of Dance & Somatic Practices,* 1(1), 99-110

Sointu, E. & Woodhead, L. (2008) 'Spirituality, Gender, and Expressive Selfhood', *Journal for the Scientific Study of Religion,* 47(2), 259-276

van Ede, Y. (2010) 'Differing Roads to Grace: Spanish and Japanese Approaches to Dance' in J. Weinhold & G. Samuel (eds.) *Ritual Dynamics and the Science of Ritual.* Harrasowitz Verlag, pp. 481-504

Jamila Rodrigues completed a PhD in Dance Anthropology and Sufism embodied ritual practice at Roehampton University. She worked as a professional dancer in Cape Town, South Africa and as a visiting lecturer at the Theology and Religious Studies Department at the University of Birmingham. Currently, she is a visiting scholar at Nichibuken, the International Centre for Japanese Studies, Kyoto. She is researching in Okinawa shamanism, indigenous knowledge and political ecology. This study explores and analyses the positive effects of shamanic praxis concerning Okinawa communities' wellbeing and explores shaman women's embodied narratives of spirit communication as a somatic process.

j_rodrigues@nichibun.ac.jp

The Poetic Body

Forget-Me-Not

Carran Waterfield

Abstract

Moving and Writing: the warp and weft of moving with words:

Through the movement of writing poetry with the body I am navigating a way through performance memory to understand and become more aware of where, how and why I have travelled this creative route these last 30 years. The body writes poetry. The body is poetic.

The Poems

Forget-Me-Not

Postures

Woman of The Golden Face

Improvisation

Moving to A Memory

Stone Ridge on The Beach

Me and Me Stick

Clinging On

These eight poems come from a place of movement. The movement precedes the words. They belong in the live moment and, at the same time, in memory and re-memory as I have moved, re-moved and re-re-moved them in finalising the collection.

I write on my feet.

I begin with **Forget-Me-Not**, an homage to my mum whose memory is fading and whose life story has driven much of my theatre work. I wrote this poem in the early stages of mum's dementia. Today I am witnessing how mum's mind slips in and out of different time frames and how the rhythm of conversation prompts her into different memory spaces as she holds, plays and re-plays her reactive memories from childhood and from her professional life in tandem. This slippage – and moving in tandem in performance terms – have been a feature of my work and explorations in body, time and space.

On this Tuesday afternoon in the lounge at her place, I find myself as interpreter explaining to her 'public' at the tea gathering that, "mum used to be a social worker" and was "brought up in a series of children's homes", in order to clarify. What she is saying might not make sense but mum has picked up on a tricky group issue in the room regarding an 'unwelcome' member of the group. She was also noting the injustice of what she perceived other people had and she hadn't: her beaker and her small cake compared with their cups, their bigger cakes, and she was really trying to do something about the problem of the 'unwelcome' person. Instinct with regard to injustice has not escaped her mind. In this instance she casts the lady sitting next to her as the one who had been left out of the group, reassuring her that everything would be okay and she hoped she "felt better now".

I am re-minded that reading signs from story, gesture and atmosphere are common to theatre makers, social workers and children. I think about memory loss and how fragmented yet totally coherent things can be if we just loosen up a bit with the meaning and let go of the logical narrative.

I am writing here by way of suggesting a navigation through these poems. While writing I am reminded of the times I have had to explain my creative work and protested that the work should speak for itself or it isn't working. I am conceding now to the need for an interpretation and like my mum, the need for a bit of an explanation.

So, you can journey with me. We will start with **Forget-Me-Not,** essentially concerning a visit to mum and a journey down to Dorset for a workshop with Sandra Reeve. Next, we arrive at **Postures**, a morning meditation at Westhay[1]. I think it speaks for itself with telling stances from Yoga practice that now inform everything I do. **Postures** prompts and ends with a memory of a 'moved' daydream that formed the core of my show **The Dig**. We journey into the past with the poem, **Woman of the Golden Face** which features the initiating side of the three-faced goddess I met in a night dream I had when I consciously chose my journey into theatre in 1986. Six years later I wrote on my feet and created **The Dig**[2]. This writing whilst not in the show underpins the entire piece.

[1] Westhay is the home of 'Move into Life with Sandra Reeve'
[2] 'The Dig' (1992). Extracts at: https://vimeo.com/88345617

Moving on from show-time to private-time we journey further back. **Improvisation** comes from the strict and rigorous performance training practice I learned with the Odin Teatret in 1990, where writing during moving was encouraged. This poem is partnered with another, **Moving To A Memory**, which catapults me back to my four-year old self in the flat in Almond Tree Avenue where my mum is the keeper of the hoover.

Flying through the years we land on a trio of poems born out of moving on Charmouth Beach: **Stone Ridge On The Beach, Me and Me Stick** and **Clinging On**, in one fell swoop. These three bring us full circle to the beginning of the collection and a journey back home through Coventry. Re-visiting them I can see they come from the gradual loss of parents and the realisation of my own mortality as I too age. They are woven with a performance memory of **Little Blue Man**[3], an homage to my dad I performed in Dorset in 2015.

These eight poems come from a place of movement. The movement precedes the words. They belong in the live moment and, at the same time, in memory and re-memory as I have moved, re-moved and re-re-moved them in finalising the collection. Sadly my mum died just as this went to print.

I write on my feet.

---

[3] 'Little Blue Man' (2015). Extracts at: https://vimeo.com/ 149065124

I am in Coventry visiting mum, on my way to Dorset:

Forget-Me-Not

I am walking up the path with mum,

in hysterics remembering when her teeth fell out
from laughing so much on the way back from work:
just usherettes staggering with laughter after getting the bus home after the
matinee.

We are walking up the path to her flat,

her stick prodding Dandelions because she can't weed anymore,
intoxicated
like that day coming home on the bus on the way back from work:
just usherettes staggering with laughter after getting the bus home after the
matinee.

"Forget-Me-Nots", she points with her stick.
"I won't Mum."

We burst again.

Two days later in Dorset:

Today's matinee is a performance of trees.
I choose the ugliest, the one with the 'been through the mill' look about it,
all churned up and inside out of it.

I like the 'pull-your-leg-up' she is giving me in her limby-winky way,
legs everywhere like Schiele's woman spreading her genitals all over the shop.

There's something about being gnawed and gnarled,
with your guts spilling out for all the children you've borne.
There's something weather worn and resilient
in this particular nearly-a-corpsed one
on this particularly wet day
at Coney's Castle where the Beeches are showing off to the 'Belles in Blue',
all lined up like tiller girls ready for curtain up.

Not my Beech.
She's a bit past it.
But's she's trying very hard still.

She mutters, "It's raining and they look a bit flat that chorus line of whipper-
*   snappers,*
just turning up in the last two weeks with celebrity on their minds.
I've been here for donkeys…
I was once a colt like them."

Suddenly out of the blue, a rural creature puts in an appearance.

An audience member!

Or,
potential participatory-co-witness-spectator,
just passing through,
beefing up his part as he keens for the loss of his favourite 'Belle Blue Show'
now a near disaster.
And this season he's especially miffed on behalf of his mates

whom he's brought over and late
for the Sunday dinner: pre-show or post-show,
depending on the fucking rain.
With the rain…the rain and "It's just such a shame," he politely howls.
The swear word is mine,
since he's treading on my patch
while I am performing like mad
in an interior kind of way:
imitating for 10,
taking in the emotion and feeling for 10,
taking on the sense of the character and being for 10,
and doing what the hell I like for another 10,
while Natasha on the bank
witness-watches me with the Ayurvedic lifestyle wisdom I read about
from a tip-off on 'Sounds True' which made me buy it on a whim,
the book, I mean,
not the wisdom.
Now I boil my water.

Back on 'Beech Catwalk' I am struck by the return of a stick-prop in my fist,
an echo of mum's Dandelion killer,
the leg-up chorusing her "I'm not dead yet" refrain again.

Then I realise it's just me watching the 'Belles in Blue'
from the point of view of a Beech and not a Yew.
Yew accepts death
because she lives forever,
well for thousands.
Beech, maybe not,
Birch, definitely not.
Mum cut her silver ones down to just stumps.

Back to the exercise:

In close-up micro-pulling-back-look
using macro with my eyes,
there's bloody thousands of the blue things
showing off,
showing up,
acting out,
acting up.
Young and springy, sprightful, wingy things taking centre stage:

The Poetic Body

upstaging me on my catwalk-turn
up the garden path with my mum in hand.
"Can I have just one last look in?"

Me, the gawky leggy one all gnarled up and battered in on the ramparts of
Coney's Castle,
working hard against the forces of the wind
and the raving rebels whose reports will slaughter me
while the 'Belles in Blue' tip tap and tap tit all over me:
cutting me up,
counting me out,
writing me off.
Breathe.

Look.

I have forgotten my own belles,
my own little patch of blues...
all wet and tattered
but still there
still there

still there...
just more raggedly spaced than those I jealously spy
from my standing point of view
on the ledge,
my roots going down while one single branch goes up,
recalling Barba's principle of opposition,
or was it Grotowski's?

Back in Coventry halfway to Southport:

After lunch
and a small white wine at The Greyhound,
me and mum carry on slowly
up the Tulipy-Dandelionless–Forget-Me-Not pathway
towards the internal moving staircase
that can turn corners.

Mum ascends
while I hum the theme tune from 'Sunday Night at the London Palladium'.
She rounds the corner on her own revolving stage
waving, smiling,
joining in with the joke.
And I say as my child-self,
"Do it again".

Not now, it's too much.
I read her thoughts
As the credits roll upwards
and time
passes
downwards.

In the early morning dew I am moving in Andrew and Sandra's garden at Westhay, Charmouth.

Postures

I go to the top of the garden
I look up – sky
I am air.

I look out – pine
I am mountain.

I look down – tulip
I am prayer.

I bend like a skier – dewdrop
I am water.

I fold down like a deckchair – twig
I am limb.

In the folding I thought for a moment it
    was road traffic,
but it was just the sea talking to me.

I stamp a gallop on paving slabs:
North South East and West.
A sign-foot-post says, "No Cars"
and the crest of an oak tree presses home the seal
of instruction and regulation.

Then I remember...

"My name's Mary,
I'm not married.
I used to be an archaeologist."

The opening line of a poem play
I made a long time ago.

I am moving with a dream and this is part of a poem play I made a long time ago.

Woman of The Golden Face

I have a recurring dream:
I am walking though a charred forest, always going down.
I pass jagged branches and a spring full of tadpoles.
A dappled blue three-legged frog sits on a rock staring:
its gullet quivering.

I walk through mud to a blue flower with star-shaped petals.
I see a path of jet black stones marked by cones and a daisy chain.
I see a charred gate fastened by rope.
It leads to a burnt out house with a plastic covering over the roof.
The door is always ajar.
I go in.
I see the back of a figure.
The head is veiled in red velvet.
The figure sits in a chair rocking and humming.

The figure turns.
She says, 'I am the woman of the golden face.
I am the one who holds the secrets of fire'.
She invites me to eat at her table of apples and
    potatoes.
She tells me they came to steal her face.
She refused.
They dragged her outside, held her face to the
    sun and set fire to her home.
It's always then that I tear off her veil and see
    that her hair is made of worms.

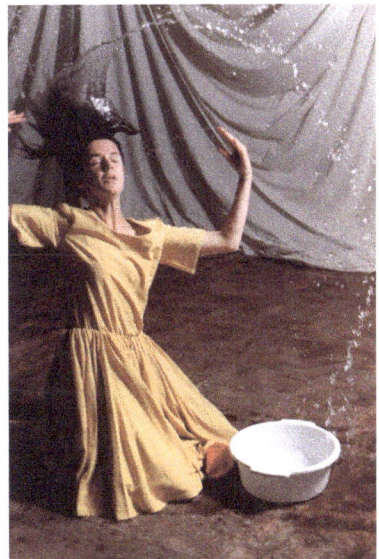

The Poetic Body

I am at Odin Teatret's November School in Holstebro, Denmark. I am working on plastiques with Norwegian actor Torgeir Wethal.

Improvisation

She stands at the bottom of the waterfall.
She looks up.
She looks down.
The pothole consumes her.
She is submerged in the belly of the earth.
She squeezes the breasts of the earth.
She is sprinkled with milk.
Her train is prepared.
Her hair is stroked.

A veil parts.
A fist thrusts, sits and smokes.
Then thrusts again.

Now the distant veil becomes a canopy,
then a large belly,
then a stone.
The stone cracks –
A peacock.
It's a boy!

And this became another poem play I made a long time ago but I said, "It's a girl!"[4] instead.

---

[4] 'My Sister My Angel' (1996), https://vimeo.com/489400834

I am working with Torgeir Wethal again and remembering the budgie in its cage in the flat in Almond Tree Avenue when I was about 4.

Moving To A Memory

It is the last day of her life.
She is placed in the cage.
She stands lightly on the hand then she
    flutters into the cage,
Wings wide then closed.
She looks in the mirror.
She swings, sitting.
She turns three hundred and sixty degrees.
She looks.
She eats.
She peers through the cage bars.
She turns three hundred and sixty degrees.
She drinks.
She looks in the mirror.
She bristles and flutters.
She goes to the toilet.
She hears the vacuum cleaner.

She darts – top right, bottom right,
Middle of the floor, top left, bottom left,
Middle of the floor, backing upstage, left corner,
Wings spread-eagled.

The vacuum sucks and she is drawn in.
It switches off.
She falls cold like a turkey on her back
Inside the Dyson, except it was a Hoover then.

I am in a pebble crater on a beach in Dorset pretending to have a bath.

Stone Ridge On The Beach

I am sat in an empty stone bath
     at the turn of the tide
         watching the waves promising to
             spill
            over
           the side
          and

drown

me

any

mi

n

u
t

e
.
.
.
.
.
.
.
.

I am on the cliff edge acting out being old.

Me and Me Stick

Me and me stick, we get along fine.
Me stick props me up.

I can triangulate with me stick
on all threes, resisting all fours just yet.
I mean, all fours or all sixes and sevens in the zimmer frame,
the frame that gets you out of the zimmer
(German for room);
out – getting aus the haus
(German)
mit schtick
(Yiddish for stück).

Stuck
(English).

Stuck in the house,
me and me schtick go out of the haus and take a look.
Aus.
Look out.
Hinaussehen:
see what's out there
(German),
not exactly but the translation sounds good.

I will fall over.
I will fall.

Me stick triangulates me out and about.
so I won't fall over.

We go in the bushes and fart in the wind.
I think about constipation.
I am stuck.
Me mud's stuck in me,
inside of me,
keeping meself in meself.

So me and me stick play stuck in the mud
with the other bronchi-cardiovascular (made up diagnosis) trees
with a dis-ease
clinging on for dear lives
just like me, me and me stick.
Us…and
Them.

I am on the clay cliffs acting out being old.

Clinging On

I am clinging on in the place of stillness, homeliness, comfortlessness,
Staying put
and looking out of the window.

From my front and from my back
peering through the windows from my house on the edge,

I am still clinging on
Pinging
Ponging on.

I see an old ping-pong ball clayed up in the foundations
from some old Christmas time cracker gone by.

Like the stone it had worn into the furniture.

Today the branch cracks off in my little brittle hand,

"Dust yourself down and venture out", she says.

I see lots of others clinging on for dear life,
so I copy them,
ape them,
empathise with them.

And we all cling on together with our fingers, hair and nails:
calcifying ourselves to the wind and the sea.

Dedicated to my movement teachers and my mum.

Acknowledgements

Torgeir Wethal and Roberta Carreri (Odin Teatret)
Ian Cameron (collaborator on The Dig and Little Blue Man)
Sandra Reeve (Move into Life)
The Dig (photographer – Damien Grazier)
All other photographs – Carran Waterfield

Carran Waterfield is an experienced theatre and performance maker, and teacher, with a body of self-devised professional work spanning thirty years. She is Artistic Director and Founder of internationally touring and multi-award-winning Triangle Theatre est. 1988 (Coventry UK). She has recently become a Sefton Borough Councillor. She is a published writer of devised performance works, poetry and creative writing. She is from Coventry and lives in Southport.

carran@carranwaterfield.co.uk ~ www.carranwaterfield.co.uk

The Proprioceptive Body

Awareness and the Question of Reliability

Laura Haughey

Abstract

Proprioception is the sense (or skill) that gives us information about the location, movement and posture of our bodies in space. It is because of proprioception that we are able to generate an intrinsic awareness of our bodies at all, as proprioception is the means by which people naturally have a pre-reflective awareness of their bodies. Many performer training methodologies seek to enhance awareness in trainees, but they don't question the integrity of the mechanism through which trainees have awareness.

This chapter demonstrates that the felt sense may not always be perceived accurately, due to the interplay between proprioception, body schema and body image, and that it is beneficial to support the integrity of the proprioceptive mechanism to ensure the reliability of the somatic feedback on which the performer relies. Working to enhance proprioceptive awareness and acuity and therefore attending to our body schema, using modalities such as wobble board training, may improve the capacity to awaken a more accurate body awareness.

B. comes into the rehearsal room with a pronounced forward bend at the neck. She complains of pain and muscle tightness around the cervical spine, travelling down to the shoulders and limiting her movement at the shoulder joints. She can't recall a particular injury – it just 'started to hurt'. I ask 'B' to straighten up as much as possible, and to retract her chin. She says that she is 'straight', and when adjusted to a more objectively 'straight' position, she exclaims that she feels 'off balance' and 'odd'. What she 'feels' is not in line with what is observable from the outside. We work with bringing her into a more easeful alignment, and to increase range of motion bilaterally at the shoulder joint... and we do it by standing her on a wobble board.

As a trainer of performers, I work with trainees who often have very little awareness of their presenting posture and physical habits. When training to be a physiotherapist, I saw first-hand that patients will often present for treatment with a verbal description of their posture that doesn't match their actual posture, as objectively assessed by the physiotherapist. Why is this? Why does their feeling sense report something very different to what an external practitioner can see? And how might recognising this 'unreliability' of bodily awareness enable practitioners to better train performers?

Body awareness is, at its simplest, the bringing of attention to the body and its felt experience, and yet this most immediate and natural of capacities can be highly elusive. As you are reading this book now, are you aware of yourself? Aware of how you are sitting and whether or not you are tapping your foot? It is probable that you were not. Drew Leder, discusses this in terms of what he calls the 'absent body':

> One's own body is rarely the thematic object of experience. When reading a book or lost in thought, my own bodily state may be the farthest thing from my awareness. (Leder, 1990: 1)

For some, much of our day, and many activities of daily living, such as walking to work, are carried out on 'autopilot', meaning that little attention is paid to the body's felt experience. As Phillip Zarrilli points out: "Stanislavsky recognised this forgetfulness of the body as a fundamental problem in both acting and everyday life" (Zarrilli, 2008: 51).

The lack of an accurate sense of bodily position is a significant problem for performers who need to fine-tune complex movements and be in command of postural and gestural languages. Indeed, the development of body awareness is a key concern for many performers and trainers of performers, as Lorna Marshall writes:

It makes sense to engage with your body, to 'invite it to the party' when you work.... It is you. Not something separate and apart. To ignore it is to ignore the fullness of yourself. To refuse a major source of information and insight. (Marshall, 2001: xiv)

Thus, the training and teaching methodologies of many performer trainers and movement practitioners seek to address the problem of the 'absent body', and to cultivate trainees' access to bodily 'information and insight'.

A first step towards addressing the 'absent body' problem is to pay attention to the body... to 'how it feels'. Bodily information (such as position sense) is already present pre-consciously, and can be raised to awareness by paying attention to it. Neurobiologist Bernard Baars uses the term 'spotlight of attention' to describe how stimuli competing for access to consciousness can be brought to attention and therefore awareness. He uses a theatre metaphor in which our senses and ideas are 'players', competing for access to the 'main stage' of conscious experience. Unconscious processes, or processes below the level of consciousness, are in the 'backstage' area (behind the scenes). We can bring different players to the conscious experience of the 'main stage', by "shining our spotlight of attention" on them (Baars, 1997: 42). For example, during a guided meditation, one may be asked to bring the spotlight of attention to the breath. Usually, very limited attention needs to be paid to the mechanism of breathing. It is a vital life process to keep us alive, yet we don't have to worry that we will stop in our sleep. Breathing is usually below consciousness, but can easily be brought to consciousness by 'shining the spotlight of attention' on it. When we bring awareness to the breath, we can slow the breathing down and control how deeply a breath can be taken. The model proposed by Baars gives us a basic understanding of the role of attention in body awareness, and yet as we have seen in the case of B, our sense of 'body awareness' can be inaccurate. In order to cultivate more objectively *reliable* bodily awareness in performers, it is important to understand the mechanisms by which we are able to bring our bodies to mind and the manner in which bodily information can be distorted.

B. points to the area of pain, around the cervical spine, and between the shoulder blades. As a performer, B. is used to working physically and expressing through her body, but she isn't aware of her habitual movement patterns, or how she is responding physically to the pain she describes in her cervical spine.

The Proprioceptive Body

A clear and consistent definition of body awareness is rarely provided in the literature, which is indicative of a general lack of understanding about how we sense ourselves (Hillier *et al.*, 2015). According to Mehling *et al.*, body awareness is comprised of both attention to, and awareness of, sensations from within the body, and relies upon the neurophysiological mechanisms of proprioception and interoception (Mehling *et al.*, 2009). We will look in more detail at these mechanisms, alongside other forms of information that contribute to body awareness, such as touch.

Proprioception is the sense that gives us information about the location, movement and posture of our bodies in space. It allows us to know where our limbs are in relation to each other and to the space around us without us having to look at them. It is the capacity that allows someone to walk in complete darkness without losing balance as they can sense and feel where their body is in space. It allows us to carry out simple and complex movement tasks and to move ourselves around and interact with the world and our surroundings. Proprioception is both the means by which people naturally have a pre-reflective awareness of their bodies, and the mechanism by which performers (and others) can develop advanced levels of bodily awareness in the service of physical skill and psychophysical fluency and expressivity.

Interoception also contributes to our 'body awareness' (ibid). It refers to the sense of the internal physiological condition of the body and includes sensations from the visceral feelings of vaso-motor activity, hunger, thirst, and other internal sensations (Craig, 2003: 500). Zarrilli defines it as our "experience of our internal viscera and organs" (Zarrilli, 2004: 658).

Proprioception and interoception are types of sensory perception in which sensory stimuli from the body are processed into neural signals by sensory receptors, and the resultant information is represented in the brain. Much of this information is pre-cognitive, but some can enter consciousness, and we can become aware of it particularly when we shine our 'spotlight of attention' on it. Mehling *et al.* therefore propose the following definition of body awareness: "the subjective, phenomenological aspect of proprioception and interoception that enters conscious awareness, and is modifiable by noted mental activities" (Mehling *et al.*, 2009: 2). This latter point is particularly important here; body awareness can be modified by our ideas, beliefs and attitudes (Pintado, 2019: 229). It is these mental activities and processes that can cause the problems of unreliability. To unpack this in more detail, let's look at how proprioception works in relation with body schema and body image.

The sensory information provided by proprioception enters into the brain via sensory pathways and is continually updated to provide us with an ongoing representation of what is happening within our bodies and around us. We can think of these representations as 'body maps'. The felt experience of the body, as represented in body maps, is known as 'body schema' (Blakeslee and Blakeslee, 2007: 32). Body schema[1] is a constantly updated map in our brains, which helps us to navigate our way through the environment and to react to whatever is going on around us. Proprioceptive information is vital to the updating of our body schema; it is a "major source of information for the maintenance and governance of movement – that is, for the normal functioning of body schema" (Ziemke et al., 2007: 279). Body schema also includes sensory information from touch and is a physiological construct created by the brain from this array of sensory information: "your brain creates it from the interaction of touch, vision, proprioception, balance, and hearing" (Blakeslee and Blakeslee, 2007: 32). It is not a static construct, but changes plastically depending on your experience as a response to what is happening around you.

Closely related to body schema, yet distinct, is the concept of body image. Body image is fed by body schema and also includes psychological factors such as beliefs, attitudes, learned behaviour and assumptions; body image is related to perception of self. Gallagher sums up the distinction between the two concepts neatly by stating, "the difference between body image and body schema is like the difference between having a perception of (or belief about, or emotional attitude towards) one's own body and having a capacity to move one's own body" (Gallagher in De Preester, 2005: 244). It is within the concept of body image that errors can occur. As Gallagher explains, "It is possible that as a set of beliefs or attitudes about the body, the body image can involve inconsistencies or contradictions" (ibid: 30).

We might expect that body schema would update body image, but this doesn't always work as it should (Blakeslee & Blakeslee, 2007). Instead, deeply ingrained beliefs about body image can be stubborn and resist being updated, leading to an incongruence between the two, resulting in the body schema (body proper) and the body image being unsynchronised:

---

[1] The term 'schema' was used as early as 1893 by neuroscientist Pierre Bonnier to "signify a spatial quality related to awareness of the body" (Gallagher, 2005: 19).

Your body schema has drifted remarkably out of touch with your body image, and you experience an internal psychic disconnect. Your body image is duelling with your body schema. Your beliefs about your body are out of sync with what your body maps or even your eyes are reporting to you.

(Blakeslee & Blakeslee, 2007: 43)

When the body schema and the body image are incongruent, it is as a result of body image and highly resistant belief networks prevent us from tuning in to proprioceptive awareness and body schemas. When we place priority on the belief networks, discrepancies between what we believe is happening with our bodies and what is actually happening can occur. Blakeslee and Blakeslee (2007) give the example of losing weight – the way our clothes fit after weight loss should give feedback to update the body schema. However, when priority is given to the stubborn belief system, this can override accurate information and the person can still 'feel' as they did before they lost the weight. In other words, whilst the proprioceptive input that we use to build our sense of body awareness and the resultant body schema are accurate, how we translate and appraise the information in regard to body image can make the resultant body awareness unreliable.

How then might movement training address the unreliability of body awareness which arises from the disjuncture of body schema and body image? One technique I use is wobble board training: a wobble board (sometimes referred to as a balance board) is a wooden or plastic, circular, flat platform mounted on a half sphere, which offers an unsteady surface designed to challenge balance and proprioception. I began working with wobble boards as a physiotherapist, and continued in my work in training performers, as research showed them to be effective in improving proprioception in the lower limbs (Waddington *et al.*, 2000). With improved proprioception comes better position sense and movement discrimination, which can in itself contribute to a more accurate sense of body awareness in these areas. But I discovered that wobble boards had more to offer than solely improving proprioception in the lower limbs: they also provide an experience to challenge our belief systems.

Along with pain, B. complains of stiffness and a reduction in range of movement at her shoulder joints. She is unaware of her excessive neck flexion at C7, which is causing a 'hump' in her cervical spine. When B. first steps onto the board, she becomes fully invested in trying to stop the edges of the board from touching the floor, and to be 'balanced'. It is challenging at first and I observe lots of saving mechanisms throughout her body, like her arms flying up, then back down to her sides as she adjusts to her body being in constant motion. On the board, I don't see the same excessive neck flexion in fixity, but lots more moving through her range of movement, as her body responds to having no fixed base of support. As she gets more used to the constant movement, and stops 'trying to balance', she settles into a more gentle dynamic journey on the board, where the sides of the board touch the floor less. Her neck is moving more freely through a wider range of motion.

Wobble board work challenges trainees by offering a different dynamic experience to their usual sensation of 'standing'. Their base of support is taken away, which challenges their balance and proprioception. Blakeslee & Blakeslee suggest this experience enables trainees to attend to their body schema[2]:

> ...the wobble board provides a powerful entry into body schema repair via stimulation of the vestibular cortex. By putting balance at the centre of attention, your body schema cannot be ignored.
>
> (Blakeslee and Blakeslee, 2007: 46)

This indirect impact on the body schema isn't a conscious mechanism. Body schema is usually below consciousness, but situations such as loss of balance, or disequilibrium between body and environment can cause the "spontaneous appearance of the body in attentive consciousness" (Gallagher, 2005: 34). The wobble board experience stimulates the body in a way that it is not used to: by challenging balance, increasing incoming proprioceptive information, and stimulating the vestibular cortex. As such, it offers a new experience in which the participant cannot do what they would normally do, since this would result in falling off the board. They must adapt, and the adjustments that they are forced

---

[2] According to Blakeslee and Blakeslee, personal trainer Jeff Della Penna also uses wobble boards, particularly when working with overweight clients, to try to bring them into "touch with their body schema" (2007: 45). His intention in using wobble boards is that his clients will be able to begin to transform their relationship with their body image.

to make can allow the sensations coming from all over the body to be brought to awareness, and not to remain 'below consciousness' as they normally would. Encouraging this information to override the stubbornly held beliefs of the body image can lessen the disjuncture between body image and body schema, lead to a more accurate sense of body awareness, and offer an enhanced understanding of postural and physical habits.

The wobble board exercises I use are varied and inspired by the trainees and the movement patterns we would like to address. When working to enhance body awareness and address habitual physical tendencies, we work quite specifically to target habitual postures and ways of moving that are limiting the trainee. Postural improvements, and enhanced awareness of how our bodies move, can be observed after just a few sessions. One exercise is movement facilitation on a wobble board, where we target certain movements, muscle groups or postural issues. These movement facilitations involve touch: a touch alone, a passive movement, or an inhibition of a muscle or muscle group to combat excessive muscular tension or strongly ingrained habitual ways of moving. This addition of touch enables another layer of information to feed into the body schema, and also allows habitual tendencies that are clear to the facilitator, but not to the trainee, to be highlighted. My work with B. was to raise her awareness of her posture and certain movement patterns that were limiting her movement repertoire and causing pain.

> After some time on the board, when B's initial 'saving mechanisms' have quietened, her neck flexion is already less pronounced, and her body is in constant and gentle motion. I then offer touch facilitation to encourage the neck to retract into a more aligned position. At first, B. completely loses balance and steps off the board. Within a few tries, she gets used to the adaptations her body makes in response to the postural change suggested by my facilitation. Her shoulders drop and the space between her shoulder blades widens. She is more aware of the physical changes happening, and can feel more movement than usual in her upper body.

The experience on a wobble board is a dynamic one; it isn't as easy to hold the body in fixity as it is in daily life. Therefore the habits that impede us day to day are challenged as the environment we are in is volatile and requires a whole new centre of gravity (which is constantly moving). As Buchanan states: "As old habitual patterns begin to dissolve, new options become possible" (Buchanan & Ulrich, 2001: 316). B. wasn't aware of her postural habit of increased forward flexion at the neck when we started working together, despite undertaking regular movement practices that encouraged the focusing of attention on her body's felt experience. B's case suggests that shining the spotlight of attention is not enough, so we needed to add in and explore the concept of body maps, and introduce B. to the idea that what feels 'normal' to her may actually be limiting her movement potential and reflecting an inaccurate body awareness. The wobble board experience enabled her to experience a habit that she wasn't aware of, and she clearly felt the release of the excess muscular tension when the habit was disrupted on the wobble board. She experienced more freedom and movement bilaterally at the shoulders, and shifted into a space of enhanced, and more objectively accurate, body awareness.

> B. reports her neck as feeling very different as she steps off the board. She spends time moving her arms from this 'new' physical posture, trying to retain the freedom in motion at the shoulder joint. She explores her new range of motion by moving around the space off the board. She is more aware of how body parts are moving in relation to each other and how this can expand her movement repertoire and potential.

This enhanced sensitivity to specific body parts and how they move in relation to each other is a result of enabling information from proprioception to update our body schema in a direct way. This can override the body image, providing a pathway to more enhanced, and more objectively accurate body awareness.

It is desirable for actors and physical performers to overcome unhelpful physical 'habits'. They can do this by cultivating a deeper understanding of body awareness through paying focused attention to proprioception, and working to lessen the disjuncture between body schema and body image by using modalities such as the wobble board. For performers, knowing that what we perceive about our bodies may not actually be accurate is useful since it enables us to be inquisitive about what we are feeling. It prevents us from making postural 'corrections' from a place of inaccuracy, which can make things worse instead of better. B's misunderstandings around her posture led her to participate in physical exercises that perpetuated and exacerbated her problems, by increasing muscle tension and limiting her movement. Now B. has a better idea about how she can train in ways that will increase the accuracy of her body awareness, and engage in movement patterns and exercises that will support her towards a healthier posture. In this way, performers can work towards making their bodies more expressive and more open to react freely to stimuli around them. They can work to eliminate 'blocks' that they may otherwise not be aware of.

Once we question the integrity of the mechanism by which we have body awareness, and recognise the interplay of proprioception, body schema and body image, we can move towards more positive modes of awakening and cultivating the types of bodily awareness desired by performers. The knowledge of how body maps interact, and the experiences on the board enable B. to move beyond solely shining a spotlight of attention on how her body feels, to the capacity to awaken a more accurate body awareness.

Acknowledgements

The author would like to thank Dr. Deborah Middleton for guidance and mentorship with this chapter.

References

Baars, B.J. (1997) *In the Theatre of Consciousness: The Workspace of the Mind*. Oxford University Press

Blakeslee, S. and Blakeslee, M. (2007) *The Body Has a Mind of its Own. How Body Maps in your Brain Help You Do (Almost) Everything Better*. Random House

Buchanan P.A. and Ulrich B.D. (2001) 'The Feldenkrais Method: A dynamic approach to changing motor behavior'. *Research Quarterly for Exercise and Sport*. 72, 315-323

Craig, A.D. (2003) 'Interoception: the sense of the physiological condition of the body'. *Current Opinion in Neurobiology*. 13, 500-505

Damasio, A. and Damasio, H. (2006) 'Minding the Body'. *Daedalus*. Summer 2006, 15-22

De Preester, H. (ed.) (2005). *Body Image and Body Schema: Interdisciplinary Perspectives on the Body*. John Benjamin

Gallagher, S. (2005) *How the Body Shapes The Mind*. Oxford University Press

Head, H. (1920) *Studies in Neurology*. Oxford University Press

Hillier, S., Immink, M. and Thewlis, D. (2015) 'Assessing Proprioception: A Systematic Review of Possibilities', *Neurorehabilitation and Neural Repair*. 29(10), 933-949

Leder, D. (1990) *The Absent Body*. Chicago University Press

Marshall, L. (2001) *The Body Speaks*. Methuen

Mehling, W.E., Gopisetty, V., Daubenmier, J., Price, C.J., Hecht, F.M. and Stewart, A. (2009) 'Body Awareness: Construct and Self-Report Measures', *PLoS ONE* 4(5) E5614, pp. 1-18

Pintado, S. (2019) 'Changes in body awareness and self-compassion in clinical psychology trainees through a mindfulness program', *Complementary Therapies in Clinical Practice*. 34, 229-234

Waddington, G., Seward, H., Wrigley, T., Lacey, N. & Adams, R. (2000) 'Comparing wobble board and jump-landing training effects on knee and ankle movement discrimination', *Journal of Science and Medicine in Sport*. 4, 449-459

Zarrilli, P.B. (2004) 'Towards a Phenomenological Model of the Actor's Embodied Modes of Experience'. *Theatre Journal*. Dec, 2004: 56(4), 653-666

_____ (2008) *Psychophysical Acting: An Intercultural Approach After Stanislavsky*. Routledge

Ziemke, T., Zlatev, J. and Frank, R.M. (2007) *Body, Language, and Mind*. Walter de Gruyter

---

Laura Haughey PhD is a theatre maker, actor trainer and artistic director of Equal Voices Arts. Laura combined her theatre and dance work with a physiotherapy degree and a year of specialised anatomy training to enhance her understanding of the body. Her PhD found the site where the two passions of arts and science were combined in the field of psychophysical performance training. Laura has taught internationally, in conservatoires, drama schools and universities, and is now based in Aotearoa New Zealand convening the Theatre Studies programme at the University of Waikato.

www.equalvoicesarts.com

The Psychodynamic Body

Dynamic Definitions for Clinical Efficacy

Sandra Kay Lauffenburger

Abstract

Body and awareness are essential components of psychotherapy, particularly when working with trauma and psychosomatic presentations. The term 'body' is insufficient if it does not capture the dynamic, flowing, physiological core of self that houses our motivations and desires and shapes our behaviours and thoughts. Equally the term 'awareness' appears too static to reflect the flowing, unfolding interactions occurring within the body. Using a clinical case to illustrate my praxis, this chapter suggests an expansion of these two concepts to reflect our dynamic human nature and to enhance clinical effectiveness.

The 'awareness' I sought for my psychotherapeutic practice was one in which the systemically unfolding interactions of the body and between bodies could be captured. I realized that I would require a dynamic, bottom-up methodology as well as a way to language these nonverbal phenomena. Accessing the nonverbal "subjective, phenomenological aspect of proprioception and interoception that enters conscious awareness, [which] is possibly modifiable by mental processes, and distinguishable from exteroception, fantasy, and thought", (Mehling *et al.*, 2011) opens a gateway to deeper self-understanding for therapist and client. The continuous flow of bottom-up data is clinically important, yet has remained outside most psychological and psychotherapeutic research.

In developing a praxis my certification in Laban-Bartenieff Movement Systems (LBMS) and extensive study of Body-Mind Centering (BMC) formed useful frameworks. The dynamic, subjectively focused ideas of these systems provide a methodological foundation through:

- exploring the ever-changing territory of the body using internal and external movement, touch, visualization, somatization, voice, art and verbal dialogue (Cohen, 2012: 1)
- understanding changes in movement qualities as reflecting changes in mind
- bringing dynamic awareness to body systems and tissues
- staying with the continual dialogue between awareness and action
- and staying in the flow of not yet knowing

A psychotherapeutic praxis also requires appropriate languaging to express and integrate inner and outer worlds. This bodymind dialogue is best suited by language which reflects the dynamic flow of subjective experience. I added this question/need to my search. Working with a psychosomatically distressed referral named Ann pushed me to evolve and deepen my understanding of body and awareness.

Introducing Ann[1]

Sitting on the couch across from me, Ann appeared friendly but said nothing. After staring vacantly at my black cat, she murmured "I have

---

[1] Aspects of this case study were previously published (Lauffenburger, 2016) to illustrate languaging the affective body. Here the case study is used to explicate my praxis.

a cat. She's old and will probably die soon." The energy of this
statement was as lifeless as its content. Saying I was interested to know
a bit more (about anything really, I thought), she talked about cats.

Ann was referred to me by a clinic specializing in mindfulness meditation for chronic conditions. Although presumably using a 'body-based' technique, their therapeutic attempts had been unsuccessful. In many body awareness practices, clients are guided to focus on sensation. Maxine Sheets-Johnstone has expressed concern that sensation is restricted in time and space, noting that "sensations are not dynamic events but punctual ones having no inherent connection or flow". She suggests turning our attention to movement which is an "unfolding dynamic event" (2010: 116). Bringing the importance of dynamic movement into clearer focus, she states:

> The living present is a matter of movement, and self-movement
> is a matter not of sensations but of dynamics. Sensations are
> temporally punctual and spatially pointillist; a push, a shove, an
> itch, a stabbing pain, a piercing sound, a flash of light and so
> on. Sensations do not and cannot give you flow…
>
> (Sheets-Johnstone, 2016: 27)

Ann, a gentle thoughtful woman in her 50s, said little and when she did, spoke in a flat, expressionless manner. She had been chronically ill for over a decade with various medically unexplained symptoms, many of which she now felt were shameful. She came to the appointment to please her worried spouse.

It took weeks before Ann related her most distressing symptom, the one that scared her practitioners the most. Fearing she would frighten me, she first revealed "the sensations are weird". As trust developed between us, Ann mentioned unbearable physiological activity occurring under the skin of her face, neck and shoulders. In time she cautiously described the wringing, wriggling, pulling, twisting, and prickling, comingled disturbingly with icy numbness that overtook her body.

Despite her vividly dynamic descriptions, the medical profession had been unable to comprehend her body's communication. Numerous medications, at one point twenty-three, were prescribed to silence her complaints. Meanwhile my body-self explored if Ann was describing deep fascial 'pulls'[2]. Nonetheless, I was more curious about their meaning for Ann. She withdrew into cold rigid silence when asked.

Since the onset of her symptoms years earlier, Ann had been increasingly worn down by the physical pain and by the attitudes of the medical profession. She had learned that to communicate with a medical professional, she had to use their language. Using words that described her true experience left them flummoxed. On one occasion early in her illness, she had tried to explain that "my cheek feels like it is dragging down into my neck and is choking me," and had been hospitalized for psychosis. After enduring many similar medicalized insults for over a decade, a wary and weary woman regularly entered my office.

Years before I met Ann, doctors decided her symptoms could be signs of palsy or stroke and insisted on multiple rounds of distressing testing. The results were inconclusive. The psychological world did not meet her either. Offering similar descriptions to previous psychotherapists Ann was diagnosed with 'repressed memories of childhood sexual abuse' and subsequently retraumatized by the therapist's need for Ann to remember the trauma(s). As a last resort the psycho-medical world proclaimed that Ann's "hysterical mind was being defensive"[3], creating psychosomatic illnesses, and referred her to me.

Many clients are unable to provide static, simple, and uncharged pain words which do not invoke anxiety and a subsequently authoritative stance in medical professionals. An ability to listen thoroughly and hear the dynamic 'bodymind' as an articulate and unified entity is not part of

---

[2] Fascia is a band or sheet of connective tissue, primarily collagen, beneath the skin that attaches, stabilises, encloses, and separates muscles and other internal organs. It is made up of fibrous connective tissue containing closely packed bundles of collagen fibres oriented in a wavy pattern parallel to the direction of pull. Fascia is consequently flexible and able to resist great unidirectional tension forces until the wavy pattern of fibres has been straightened out by the pulling force. Fascia becomes important clinically when it loses stiffness, becomes too stiff or has decreased shearing ability. Inflammatory fasciitis or trauma can cause fibrosis and adhesions, and fascial tissue fails to differentiate from the adjacent structures effectively. When fascia is tight and distorted it can torque, pull, jam, and compress the surrounding nerves, muscles, bones, joints, skin, and blood vessels, causing misalignment, imbalances, pain and dysfunction. Vicariously entering Ann's experience, I felt her descriptions were most similar to what traumatised, adhesed fascia might feel like. However, her experience was ever-changing and migratory.

[3] Direct quote from Ann in a session.

medical or psychological training. Patients are more often dissected into medicate-able symptoms as in the case of Ann.

Additionally, patients despair as they become caught in biomedical discourses which question the legitimacy of their experience, disempower their search for appropriate treatment, and place blame on them for suffering (Lauffenburger, 2009). If like Ann, the client describes subjective internal (and dynamic!) experiences, 'demonization' as alien, dangerous, or crazy may occur (Cates, 2014; Lauffenburger, 2016). Such experiences had occurred regularly for Ann and had rendered her mostly mute.

> *In mutual muteness Ann and I met twice weekly. Although saying little, I found Ann non-verbally articulate. I would observe Ann trying to sit primly on the couch. Sooner or later this posture would segue to squirming and wriggling. Awkwardly she would shift position, rearranging pillows and self every few minutes. Neither her inner nor outer world appeared to offer comfort. Often Ann would suddenly topple onto her side and lie there, unable to sit up.*

> *Outside of sessions I motorically entered Ann's world, replaying her movement sequence with my complete bodymind. Feeling the constant readjustments of disjointed body parts, abrupt shifts, and jerky muscular contractions, my imagination associated to the unsettled discomfort of an infant whose basic needs, such as nappy change, might have gone unnoticed, forgotten, or worse yet, neglected. Replaying Ann as she tumbled onto her side, I felt myself giving up. Trying her side-lying qualities, I sensed into her comment "I am exhausted". To me the side-lying feeling and these words did not match. In my physical replay, I felt disconnected rather than tired.*

Self Psychologically-trained[4], I had learned to vicariously introspect[5] to attune to the client's emotions and organizing principles, and then verbally

---

[4] Self Psychology is a form of psychodynamic psychotherapy that posits that an individual's self-cohesion, self-esteem, and vitality derive from, and are maintained by, the attuned responsiveness of others to his/her needs. In Self Psychology, the therapist makes the effort to understand the client from within the client's own subjective experience and viewpoint.

[5] Vicarious introspection is a term introduced by Heinz Kohut to describe how a psychoanalyst should approach empathic understanding. Many therapists see empathy as placing themselves in another's shoes and looking outward. Kohut (1981) saw that experience as 'extrospection'. Vicarious introspection is placing oneself in

reflect my understanding. Working with Ann in this manner was unsuccessful. Verbal reflections were met with an empty stare. Our therapeutic world felt lifeless.

> In an early session, with worry and concern, Ann asked: "How will you work with me?" I stuttered… "my hope is to do my best to understand your feelings…" But at the word 'feelings' she stopped me and, warding me off as if I were a vampire, flatly said, "No. Not feelings. Feelings are dangerous." She then returned to her silent squirming sitting.

At that point I realized my only connection to Ann was through being more deeply aware of myself. Most psychotherapists agree that their own embodiment provides observational skills to track the client's responses (Dosamentes-Beaudry, 2007). However, attention is typically placed on shifts in posture or repetitive gestures (Sletvold, 2014). I needed to delve somatically deeper into myself and hopefully connect to Ann.

During a Daniel Siegel workshop discussing the Wheel of Awareness (Siegel, 2012: F-10) a continuous awareness process called 'streaming'[6] (Kepner, 2015: 601) was introduced. Streaming focuses on the unceasing flow of interoception and offers a dynamic body awareness methodology. In my personal explorations, I initially discovered that streaming the entire body was insufficiently refined for intersubjective psycho-therapeutic work. Working with my own sensory experience, I played with more specific processes such as 'listening with the skin' or 'taking in light through the pores', and other system-specific foci I had learned through BMC.

Using BMC somatic awareness processes combined with streaming, I had plumbed my physiological systems. Now I applied my discoveries: in a therapy session with Ann, I streamed my skeleton system, naming it to myself as 'sitting in my bones'. I realized this was a familiar somatic place for me; I stream it regularly when working with most clients.

---

another's shoes and looking inward, as to the effect of being in 'those shoes' on one's affective response and sense of self.

[6] Streaming was initially defined as a continuous flow of energy (Kepner, 2015). In Siegel's use it is a practice similar to internet usage where it refers to the process of delivering or obtaining media continuously without downloading the entire file.

My bones' weighty clarity offers a reassuring place in which to ground myself as I attune to clients. It is a home base from which I access a second level of awareness emerging from clients' bodies. Neurological research suggests that mirror neurons pick up more than external movement, that somatic information may also be conveyed (Ernst *et.al.*, 2012). Sitting in my bones heightened my ability to differentiate my own feelings from my client's feelings, while reciprocally offering clients a grounded experience to mirror and to incorporate into their sense of self.

Unfortunately, sitting in my bones with Ann offered little. All I could feel was myself, which I found atypical. I decided to 'sit in my (intra and intercellular) fluids' instead. I know the streaming nature of my fluids intimately; they 'buzz', 'zip' and 'zing' with aliveness. My fluids are weighty but with a different dynamic than the bones' weightiness. My bones lend me clarity, direction, and solid purposefulness while the fluids provide a sense of continuity, flow, and vitality.

> *The next session, while sitting in my fluids with Ann, I experienced a gradual dimming of vitality. I looked at the body across from me and saw a wide-eyed, empty stare and tense, non-breathing rigidity. The room lost its oxygen. My mind named this as 'fading away'. I asked Ann if she noticed anything. She shrugged and stated matter-of-factly "Oh, I often become zombie-fied".*

The dual awareness, of my own fluid physiology and a foreign 'evaporating' one allowed my first deep insight into Ann's inner world. A zombie is a fictional undead being created through a partial reanimation of a human corpse. As humans we are horrified to imagine the 'living dead', a body container that walks, searches, and tries to incorporate the vitality of others by eating them. This idea, and the Hollywood depiction, engender terror.

> *I reflected on how Ann terrified the medical profession, their interventions with strong and sometimes debilitating medications, and their seeming inability to empathize with her experience. Was their reluctance a counter-transferential enactment of their own disavowal of death? I wondered why I didn't feel fear. Did my well-honed body awareness allow me to separate her 'deathly' energetic flow from mine? Was it my ability to be curious about her somatic body experiences, rather than avoiding them? On a deeper and*

existential level, I wondered if the previous professionals (as well as family who also shunned or shamed Ann) unconsciously sensed 'death', could not allow it into consciousness, and 'annihilated' Ann.

The 'fading' discontinuity of Ann's presence did not match my understanding of the psychological concept of dissociation. Letting go of the label, I deepened into Ann's internal experiences. When a professional uses a psychotherapeutic label, such as anxiety, sexual abuse, or even dissociation, fuller understanding of the client's experience is closed down. Labels mislead or distract client and therapist from more useful information. For example, when a client is assigned the diagnosis of anxiety by their GP or psychologist, both parties assume they 'have the answer'. In truth, they have only stopped the questions. Unpacking the sensations and motor impulses lumped into and lost within the word 'anxiety' can reveal a rich and diverse array of subjective experience(s) unique to the individual. The diagnostic label erases emotional motivations, kinaesthetic information, and deeper truths found within the dynamic feeling self. I saw my therapeutic work as exploring *inside* Ann's experience to access its nuanced richness and to discover her nuclear self[7].

Ann repeatedly said "I have no memory of my childhood. I barely remember my 20s except for what my mother tells me." Nonetheless the streaming process allowed me to reflect on a number of psychological scenarios, such as possible attachment experiences[8]. Questions about the physical and emotional availability of childhood caretakers emerged. Were they present only fleetingly? Did they waft in and out of her young life, leaving her with a discontinuous psycho-physical-emotional self-reflection and an inconsistent sense of presence? Had she learned to become invisible or did she need to be

---

[7] Nuclear self is a Self Psychological postulate (Lee, Rountree & McMahon, 2009) referring to inborn positive growth potentials dependent on appropriate and relational environmental response.

[8] Attachment history was only one possible hypothesis offered by the interoceptive data. I also reflected on the LBMS Effort component of weight, particularly the 'floaty weightlessness' I observed as Ann 'faded away'. I considered its possible connection to her lack of agency, or inability to make an impact upon the work, including not being heard. Frank suggests "the ability to experience body weight is forged in early infancy through the emergence of developmental movement patterns that aid the ability to feel here and in clear relationship with the body self" (2001, p.76).

invisible for them? Perhaps after all her medical experiences of having bodily dynamics dismissed, devalued and shamed, I was experiencing a more recent protective strategy?

Dual dynamic interoceptive awareness allowed our work together to advance from a bottom-up perspective. I captured key internal moments to dynamically offer back to Ann. She was intrigued. Never before had anyone listened so deeply to her. No one had tried to understand her body's communications without shaming or silencing her. Ann grew curious about her body's communications and asked to learn streaming. We entered a mutual fascination with the aliveness inside her.

> *Trusting me more, Ann wanted to share her inner world. I introduced present participle 'verbing'. I imagined Ann would offer me verbs like 'buzzing' or 'whirring', etc. Hearing my suggestion, Ann balked. The English language felt too restrictive for the marvellous things she was experiencing.*

Capturing the systemically unfolding interactions of the body was my intent, and 'verbing' was a process I learned decades earlier during my training as a Certified Laban Movement Analyst. I thought it was a great idea, but, vicariously introspecting, I realized that from Ann's point of view, I too might be imposing *my* language needs on her. I began to understand how I was mirroring the medical world's inability to let her speak. I had to find a way to make room for her language.

> *With child-like delight, Ann and I extended verbing to include nonsense or self-made-up words. Loving this playfulness, Ann regularly created onomatopoetic words. My favourite was "IRK!-spackling", which she felt embodied the sudden disruption of her internal flow when she felt threatened, and the subsequent jerky, sparky crackling that happened as her inner flow began reorganizing.*

In psychotherapy, I believe we do not develop awareness to facilitate change; awareness *is* the process of change[9]. As Ann's interest in her inner world expanded, the zombie-fied wispy ghost began transforming. Things

---

[9] Perhaps awareness should be a present participle ('awarenessing') in order to acknowledge the process of changing.

were happening inside her, which I believe directly resulted from our enlivening, DYNAMIC interactions.

> A bit over a year into the therapy, rather than just toppling over, Ann asked "would it be ok if I lie down? My face [the pulls] is exhausting." Pleased she could state her need, I nodded. Ann lay quietly and slept for 20 minutes. After napping, Ann sat up, excitedly exclaiming "my cheeks and mouth are bluckle-ing!" Using her hand, Ann showed me the felt-action: rhythmically opening/unfolding and closing her fingers and thumb, and telling me it felt like a flower blossoming.

I personally (and to myself) associated her action to the suck/release rhythm of an infant's suckling reflexes. Bluckle-ing (which I imagined to be something like 'blooming' and 'suckling') was Ann's 'verbing' of her emerging vitality. Both Ann and I could feel that she was returning to life. Ann continued seeing me for a number of years after this point, and over time we both would reflect on the personal transformation that had begun to manifest in this session.

Evolving Praxis

Exploring the dynamic interoceptive body through streaming awareness provided several working hypotheses:

1. Awareness is best conceived as having a developmental and relational trajectory
2. The body is animated physiology, a 'morphology of physiological systems in motion' and emotion (Sheets-Johnstone, 2010: 117)
3. Languaging of body awareness must honour the experience of the client rather than use the vocabulary of the professional. Creative freedom in languaging must be encouraged or taught
4. Awareness of the body must capture the ongoing, dynamic nature of the human body

Mehling *et al.* (2011) formulate a developmental progression for body/self-awareness. Based on his qualitative research and reading of Gadow's (1980) phenomenology, Mehling identified four levels in the actualization of body-self-awareness, progressing from:

1. The lived body state – where the body is taken for granted and patients are unself-consciously aware or unaware of it. Many

people live their daily life here, until trauma, injury, or pain, create attention to the body.

2. The objective body state – where the body is experienced as opposing, in tension or disunity with the psychological self. In this state, patients present their symptoms for treatment.
3. The cultivated immediate state – where a new relationship characterized as acceptance of body experiences occurs. Objectification decreases and attempts to make sense of the experiences begin.
4. The subjective body state – where the body is experienced with curiosity as a source of learning and meaning, endowed with intelligence. The body no longer constrains or limits but becomes an integral aspect of self and a locus of consciousness. (Mehling *et al.*, 2011: 10).

My clinical work requires a relational component to be included in the definition of awareness. Phenomenological philosophers propose a body awareness/self-awareness dialectic, but Mehling hypothesized a 'trialectic' between body-awareness, self-awareness, and environmental/other-awareness (ibid: 10). By extension, in clinical situations I suggest a dual trialectic occurs, where the client's trialectic intersubjectively interacts with the therapist's trialectic. The client's developmental progression of awareness depends on the therapist's developmental state. Hopefully, the therapist has grown in body/self and relational awareness significantly more than the client, so the client can draw on the therapist's psycho-physical-emotional functioning. This parallels the selfobject construct of Self Psychology (Lee, Rountree & McMahon, 2009) where the client 'uses' the therapist's better developed psycho-emotional functions until the client develops them for herself. It also echoes the intersubjective infant-caretaker interactions in infant research where the infant relies on the caretaker to provide psycho-emotional and physical functioning until the infant develops sufficiently.

Conclusion: Definitions of Awarenessing, Morphologies-in-Motion and Languaging

I first propose a definition of awareness constructed dynamically of developmental process, a trimodal structure (body-self-environment/other), and a relational modifier.

Secondly, the definition of body must be more dynamic for effective clinical practice. First used in the 1300s, the word 'body' signified corpse or container. This inert form is not what Ann and I experienced. We were dynamically interacting physiological processes bounded by a morphology that contained and adapted to these processes. Sheets-Johnstone uses the phrase 'morphologies-in-motion' to emphasize the animated, moving (internally and externally) nature of the container/contents. Using 'morphologies-in-motion' instead of 'body', the body-self-environment is implicitly placed in relationship, and incorporates somatic information:

> Morphologies-in-motion are first and foremost subject-world relationships…[living] not in a vacuum or in ambiguity but in a world particularized for it by its being the animate form it is. Precisely because it does not live in a vacuum or in ambiguity, it is unnecessary to 'embed' it in a world, just as it is unnecessary to 'embody' its actions, cognitions, experiences, emotions, and so on. (Sheets-Johnstone, 2016: 8-9).

As demonstrated with Ann, working with the internal movement of one's physiology was clinically effective. Through a non-static approach, the inner life of those who find it difficult to be with themselves, or who have been muted, becomes available.

Finally, languaging is basic to being human. Self Psychology notes three critical motivators of human behaviours: to see and be seen, to understand and be understood, and to belong (Lee, 2001, pers. comm.). The need to understand and be understood includes the body-self-other connection, and requires language for full integration. To gain/regain a body-self-other connection, a client's developmental awareness progression must include a process of languaging *that is their own*, and not imposed by others. Clinically, I have found that disruption or inhibition of a person's languaging can be as traumatic as disruption/inhibition of their body-self-awareness. For Ann, both of these were affected.

Language must capture dynamic human experience. Lakoff and Johnson (1999) suggest this is the role of metaphor, but I propose metaphor must be dynamic. The interoceptive experience of our morphology-in-motion is rich and highly informative. It has always been key to being human as well as the essence of health and wellbeing. Approaching our experience with curiosity, playfulness, and linguistic freedom, affording it dynamic energy and representing it through dynamic metaphor, present participles, unfolding flowers, and 'IRK!-spackling innards', transformation occurs.

References

Cates, L. (2014) 'Insidious emotional trauma: The body remembers...', *International Journal of Psychoanalytic Self Psychology*, 9, 35-53

Cohen, B. (2012) *Sensing, Feeling and Action: The Experiential Anatomy of Body-Mind Centering*. Contact Editions

Dosamantes-Beaudry, I. (2007) 'Somatic transference and counter-transference in psychoanalytic intersubjective dance/movement therapy', *American Journal of Dance Therapy*, 29(2), 73-89

Ernst, J., Northoff, G., Böker, H., Seifritz, E. & Grimm, S. (2012) 'Interoceptive Awareness Enhances Neural Activity during Empathy', *Human Brain Mapping*, https://doi.org/10.1002/hbm.22014

Frank, R. (2001) *Body of Awareness: A Somatic and Developmental Approach to Psychotherapy*. Gestalt Press

Gadow, S. (1980) 'Body and Self: A dialectic', *The Journal of Medicine and Philosophy*, 5(3), 172-185

Kepner, J. (2015) 'Energy and the nervous system in embodied experience' in G. Marlock & H. Weiss (eds.). *The Handbook of Body Psychotherapy and Somatic Psychology*. North Atlantic Books, 600-614

Kohut, H. (1981) 'Introspection, Empathy, and the Circle of Mental Health' in P. Ornstein (ed.) *The Search for the Self, Vol 4*, International Universities Press, pp. 537-567

Lakoff, G & Johnson, M. (1999) *Philosophy in the Flesh: The Embodied Mind and its Challenge to Western Thought*. Basic Books

Lauffenburger, S. (2009). *Problematizing Chronic Pain Treatment: A Qualitative Study Using Foucauldian-Informed Discourse Analysis*. Unpublished thesis submitted to Charles Sturt University, B. of Social Science (Psychology Honours) degree

_____ (2016) 'Demonized body, demonized feelings: Languaging the affective body', *Journal of Body, Movement and Dance in Psychotherapy*, 11(4), 263-276

Lee, R., Rountree, A. & McMahon, S. (2009). *Five Kohutian Postulates: Psychotherapy Theory from an Empathic Perspective*. Jason Aronson

Mehling, W., Wrubel, J., Daubenmier, J., Price, D., Kerr, C., Silow, T., Gopisetty, V. & Stewart, A. (2011) 'Body Awareness: A phenomenological inquiry into the common ground of mind-body therapies'. *Philosophy, Ethics, and Humanities in Medicine*, 6, 1-12

Moore, C. (2014) *Meaning in Motion: Introducing Laban Movement Analysis*. MoveScape Centre

Sheets-Johnstone, M. (2010). 'Kinesthetic Experience: Understanding movement inside and out', *Journal of Body, Movement and Dance in Psychotherapy*, 5(2), 111-129

_____ (2016) *Insides and Outsides: Interdisciplinary Perspectives on Animate Nature*. Imprint Academic

Siegel, D. (2012) *Pocket Guide to Interpersonal Neurobiology: An Integrative Handbook of the Mind*. W.W. Norton

Sletvold, J. (2014) *The Embodied Analyst: From Freud and Reich to Relationality*. Routledge

---

Sandra Kay Lauffenburger, B.Ed. M.Sc. B.Soc.Sci. (Hons Psych), Grad Dip (Adult Psychotherapy), Dip (Dance Movement Therapy), Registered Clinical Psychodynamic Psychotherapist PACFA 20209, Registered Dance Movement Therapist DTAA (DMT Prof.) 203-01, Certified Laban Movement Analyst LIMS maintains a clinical practice in Psychodynamic Psychoanalysis and Dance Movement Psychotherapy (DMP) in Canberra, working with a spectrum of issues from complex trauma through to psychosomatic issues. Over 35 years of exploring body and movement therapies as well as clinical training in psychoanalytic Self Psychology inform her work. She is an invited presenter on applications of LBMS for DMP in Australia, Canada, and Asia. She has taught LBMS, DMP, and psycho-dynamics for the Wesley Institute (Sydney), IDTIA (Melbourne), University of Auckland (NZ), and Apollo Training Institute (Beijing). She is currently President of the Dance Movement Therapy Association of Australasia.

slauf@netspeed.com.au ~ www.selfnmotion.com.au

The Signing Body

Sheila Ryan

Abstract

Signs and signals are common to all forms of organic matter. Biosemiotics is the study of non-linguistic signs and signals in the biological world. The system of medicine called homeopathy is an example of applied biosemiotics. This chapter describes the challenge for homeopathy of observing and interpreting these elemental signs and signals. We are all storytellers and it is difficult to perceive the underlying sign pattern beneath the human story. It is only at this level however that the connection between human suffering and the natural world of remedy substances can be made. In following the clinical case of a child who does not walk, the story of the case is gradually listened to at the level of the sign beneath the human story.

De wo tan a hafiz a na
One crawls before walking
(Ewe proverb)

Woezo, Efoa? Welcome, how are you?
Me fo. I am fine.

Wo ha efoa? And how are you?
Me fo.

We sit awhile and talk of local matters. We watch and we wait.

In Ewe culture it is important to observe the rituals of greeting, to take time to come to the point, to leave space for sitting together in silence, to allow the story to unfold in its own way and time. The mother eventually begins to tell us what the matter is:

> "Esianyo refuses to leave the breast and to move around. She must be carried everywhere. She cannot walk although her 'milk brother' has been running around for six months now."

We see Esianyo's legs wrap around the waist of the mother as she clings to the breast. When she is set upon the ground, she cries and reaches up, her thin, wasted legs holding the same encircled form; her thighs spread, her feet close together, soles facing each other.

Esianyo is 18 months old. Her adoptive mother brings her to the clinic for help with weaning. I am here as a member of the Ghana Homeopathy Project at a village clinic in the Volta region of South East Ghana to assist Emperor (Samuel Tsamenyi), the chief clinician, in practising homeopathy. At this 24/7 clinic, homeopaths work alongside nurses and midwives helping people through life, from birthing to dying. As a low-cost, low-risk alternative to the largely unavailable allopathic medical support, homeopathy has been asked for by this village clinic and welcomed by the Ghanaian government TAMD (Traditional and Alternative Medicine Directorate). This system of medicine uses non-toxic micro doses that would in larger amounts produce in healthy persons symptoms and signs similar to the disease.

In this chapter I have intertwined stories: that of Esianyo, the patient, the story of 'the signing body', the practitioners' story and finally the stories of the remedies prescribed. So, are you listening *with greased ears,* as the old storytellers in Esianyo's village would have said? Then let us begin at the beginning, as all good stories do.

Once upon a time, over 3 billion years ago, the anaerobic earth's atmosphere gave birth to single-celled bodies. Some of these, the Methanogenic Archea, for example, live on to this day in our human guts. For their incredible feat of survival, these highly adaptive microscopic bodies depend upon giving and receiving chemical signals.

We now know that all life forms are continually moving, adapting and re-forming themselves in relationship to each other and the changing environment, using signals: chemical signals evidenced by colour, odour, temperature, gas and precipitate changes and reactive signals in moving away and towards, opening and closing, separating and merging. Signal here refers to the reaction or response itself. Sign can be seen as evidence of this reaction – as in, for example, signs and symptoms of disease.

Biosemiotics meaning, literally, 'life signs' studies the meaning of these pre-linguistic signals and signs within the context of the biological world. Most bodies in the world do not have brains. The supplementary organ of the brain comes late in evolution and to only a few species. Yet signalling and signing is common to all organic life from the single-celled organism to the most complex being. The single-celled protozoa, the amoeba, for example, senses chemical changes in its environment and moves in response by continually changing its body shape, forming temporary 'false feet' from its cytoplasm into which its body then flows. Some shelled sea creatures, like the crabs, shrimp, lobster and pencil urchins, are building up more shell in response to the acidification of the sea due to rising levels of carbon dioxide. These are signs of climate change.

Homeopathy can be seen as an example of applied biosemiotics in the medical realm. Instead of suppressing signs and symptoms of disease the homeopathic process is to observe them in order to perceive the unique pattern of dis-equilibrium in each individual case. The homeopath then looks to the natural world for a non-human organism displaying a similar pattern. When administered in micro doses, a remedy made from these non-human biological organisms, according to the principle of 'like cures like,' effects a restoration of equilibrium: that is, restores freedom to move, to react and respond.

The poisonous plant *Atropa belladonna*, commonly known as Deadly Nightshade, for example, disrupts the parasympathetic nervous system affecting involuntary activities like breathing, heart rate, balance and dilation of blood vessels. Dilatation is its principal action. Taken in potency and according to homeopathic principle, it is a curative remedy for people, notably children, with high fever, delirium, dilated pupils and flushed face. A substance is potentised through a process of serial dilution and succussion (shaking). The power of medicines prescribed in potency was discovered by the physician Samuel Hahnemann (1755-1843) during experiments to reduce their toxicity. Potency is currently explained in theory by developments in quantum physics. We can view this 'signing body' as a quality of the 'ecological body':

> Body and environment, I suggest, co-create each other through mutual influence and interactional shaping...The ecological body is situated in movement itself and as a system dancing within systems rather than as an isolated unit.
>
> (Reeve, 2011: 48)

In a state of disease a body can start to isolate itself in the sense of no longer being part of a fluid, adaptive signing process. The body shows signs of signalling in a repeated, patterned way, without reference to the changing environment. For example, rising body temperature in a warm room signals time to remove a layer of clothing or open a window but continued raised temperature whatever the environment signals a loss of homeostasis, that is dynamic equilibrium.

In order to maintain equilibrium within the 'dance' we therefore tend to resist change, to hold onto our shape even while doing so puts us at risk of fixing ourselves in an unresponsive pattern. *We* are paradoxically both bodies in motion and bodies resisting motion.

As I pay attention to Esianyo I am reflecting that if she cannot be helped to move from her clinging posture with her legs wrapped around her mother's waist, she is at risk of weak legs and poor motor skills when she does eventually try to stand and walk. What is the matter with Esianyo? What disequilibrium is her a posture a sign of? And what is the matter, quite literally, the similar animal, mineral or plant matter, which answers her sign?

Receiving the homeopathic case is an emergent rather than a reductionist understanding of healing and diseasing. We can say that the signing body of the homeopath perceives the signing body of the patient within a process of participant observation. In this way, the homeopathic receiving of the case mirrors the natural world of interactive co-creation.

Body and Awareness

Esianyo's adoptive mother tells us the history of the child and in so doing helps create an understanding of the context in which Esianyo's particular fixed posture is co-created in dynamic relationship both to the events in her young life and to her own nature and predisposition:

"Esianyo was born to a mother who went to her local prayer camp to give birth instead of attending the village outreach clinic where traditional birth attendants are available. Her birth mother died from postpartum haemorrhage. I had given birth at about the same time and I elected to raise the child. The sound of keening from mourning women surrounded the hungry baby. Esianyo was very distressed by the time I reached her. Now, 18 months later, she shows no inclination to get down and leave the comfort of her carrying cloth."

As the clinic is conducted in Ewe language and translated into English, the importance of participant observation and sensory perception is heightened. We human beings are all story makers. Story is how we make sense of our lives and story is central to the now fading oral culture of the rural Ewe. Here, personal, cultural and social values are taught through proverb, fable, song and ritual practices. In receiving village cases, we observe rituals of greeting and exercise patience in watching and waiting for the story to unfold.

The key is to listen to the story as part of the pattern being expressed by the signing body. For example, people coming to the rural clinic would say,

"It hurts here (pointing to the heart) like a snake coiled tight round"

or

"I run like a startled chicken into the bush when I hear gunshot"

or as in the case of Esianyo, her mother says:

"She is like a stone, she refuses to move."

These words spoken spontaneously, especially if repeated and accompanied by gestures, turn a *story* into a case upon which a similar remedy can be prescribed. "Metaphor is the moving line" (Ryan, 2004: 114.) is how I characterise the importance of this. A metaphor, spontaneously formed, carries a liveliness, a direct connection with form and motion. It is most often an indication of the motive force behind a story. What is the quality of Esianyo's fixity? It is obdurate, grounded, weighted, like a stone.

And did Emperor and I listen to the story in this way? Well, we shall see. Are your ears still greased?

No, in fact, we did not at this point pay sufficient attention to the elements of form, posture, gesture, fixity and movement. We did not squint our eyes and settle down in stillness to gaze at Esianyo, to perceive her metaphor, to perceive that she was 'as if' something else. We were sad to hear that the birth mother went to the prayer camp to give birth instead of to the clinic. We were sad to learn that as a result she bled out and died. We were frightened and angered by the number of women we hear following the evangelical priests in all matters including their health and wellbeing. We listened and we interpreted Esianyo's case in the light of our own responses to her story. As a result, the first remedy prescribed, Natrum Muriaticum, was based upon the interpreted and projected emotional state of the child rather than on observation and perception of her patterned form and motion.

This remedy is derived from sea salt and can answer the state of 'never well since grief and loss'. It is one of the remedies indicated in developmental delay and appeared to answer Esianyo's state. The nature of its characteristic posture is expressed by the tale of Lot's wife who looked back and as a result was turned into a pillar of salt. Natrum Muriaticum preserves grief, seeing all life through a lens of loss. There it is, the gesture of looking back and the resistance to movement, to change, in a futile attempt to regain homeostasis. We were 'seduced' by the all too human storytelling and so missed the point as we shall see…

The homeopath continues with receiving the case until the gestalt of the answering homeopathic remedy is revealed. Until they can answer the question 'What 'non-human specific' state, as Rajan Sankaran calls it, does the signing body express?' The task is to perceive through a quality of looking, without fear or favour (Ryan, 2004), the repeated signals, signs and symptoms, gestures, original language, metaphor, attraction, repulsion, in/stability, pace, sensitivity, cravings, aversions and modalities.

For example, is the person as if the windblown flower, Pulsatilla, changeable, irresolute and as thirstless as the dry soil that flower prefers? Or are they exhibiting the clawing gesture of the Black Panther and experiencing a "sensation as if coiled up and ready to strike" so strong is the "sense of being under attack?"[1]

---

[1] These are sensations which were elicited during the homeopathic proving of the Black Panther remedy. (Sankaran, 2014).

We signal distress and develop symptoms and signs of dis-ease but very often we ourselves cannot move out of these patterns without help, without an answering gesture. A homeopathically prescribed remedy is one such gesture.

The mother returned to the clinic a month later, Esianyo still in the carrying cloth. She was sleeping better, we heard, no longer waking so often in the night to feed, but otherwise no change.

This time we observed more closely the quality of Esianyo's form and motion. We touched and gently pressed her thin, wasted legs. We felt their brittleness, as if they would break if forced from their encircled position. We noticed how fixedly she held them in the same shape when her mother sat her upon the ground. When she stretched up with her arms but made no attempt to move her legs to get nearer to her mother, we noticed how little awareness she seemed to have of her legs. We tested this in waggling her feet to encourage her to use them to reach her mother. She looked at us and shifted her feet only to return them to their position when we left them alone. (It is the norm in the villages for children to be breast fed as toddlers, but not to delay crawling and walking in order to cling onto the breast.) We began to perceive (by means of sensory observation) rather than to interpret (by listening to the story), the meaning of Esianyo's signing body.

"Ah ha!," we said to each other, "it is as if she might break, like glass". Our sense of Esianyo, that she is brittle, like glass, again exemplifies metaphor as a moving line. She is not just any mineral, any stone. She is a brittle one, like glass. We are refining our sense of her through the 'as if' imagery of metaphor.

The remedy Silica is derived from the chemical compound silicon dioxide, one of whose forms is quartz. Silica is the most abundant mineral found in the crust of the earth. Characterised by its hardness, it is the principal component of most types of glass and other hard substances, for example, concrete and in more recent history, microchips. These manufactured uses show its quality of maintaining shape and being inflexible. It can break but not bend. It stands hard and keeps its shape at any cost.

It is only quite recently that scientifically we begin to understand the role of this micronutrient in maintaining good bone health in humans.

However the homeopathic potentised Silica has shown its affinity to growing bones. since it was first proved[2] by Hahnemann in 1828.

Esianyo's characteristic non-human specific posture and movement is like the inflexible shape and form of quartz, even to the sense of her legs being brittle and more likely to break than bend. Silica (like the previous Natrum Muriaticum) is an element indicated in developmental delay, when the signs agree. For example, on further enquiry, we discovered that Esianyo tends to be constipated, straining even with soft stools. This is characteristic of children in need of homeopathic Silica. She has the typical swollen abdomen and thin legs of the Silica child, although this is a look too common among the village children to be characteristic. We observed that when she is set down on the ground she will eventually settle down to play, scratching in the dirt with her fingers, the tiny debris she finds there seeming to hold her attention. In the Materia Medica, Silica is described as having fixed ideas, fixed on tiny things, like pins, ordering and counting them. (Phatak, 1988: 541-546)

> *We prescribed the remedy Silica in potency 30 and slowly, over time, Esianyo relaxed her posture and became stronger in the legs. She didn't crawl but gradually, over the next two months, she stood, held on with her hands, wobbled and then took her first steps.*

The Ewe proverb at the beginning of this chapter reminds us that in the normal, healthy course of things, crawling comes before walking. For Esianyo, this natural order was interrupted, perhaps by the poor nutrient status of her birth mother, perhaps by her loss at birth. We don't know. We can only pay attention to the signing body of the client in dynamic relation to the signing body of the practitioner and then look to the natural world for a similar, answering, pattern.

Esianyo's name means 'All is well' in English. In being adopted by such a patient and observant woman who then brought her to the homeopathic clinic, she is indeed well named. And so we come to the end of our story.

---

[2] In a proving a homeopathically prepared substance is administered to healthy volunteers in order to produce the symptoms specific to that substance and thereby reveal its inherent curative powers. The effects which occur are documented and systematically arranged to form a symptom pattern or 'remedy picture' which is specific for that particular substance. Provings are always conducted at a non-toxic level, i.e. by using substances with a sufficient degree of dilution to guarantee the safety of the medicinal product. (The European Committee for Homeopathy, 2005)

And have you been listening with greased ears? Then may your tongue be oiled so that you in turn may pass on this story.

Acknowledgments

With thanks to Samuel Komla Tsameny and patients of the Mafi Seva Clinic who gave permission to share their cases. Thanks to participants in Sea Change Supervision workshops for their enquiries into mindful and embodied observation. Thanks to Sandra Reeve's Move Into Life workshops for helping me to become more aware of habitual movement, and through moving in the landscape, to make deeper personal connection with the non-human.

References

Barbieri, M. (ed.) (2007) *Introduction to Biosemiotics: The New Biological Synthesis.* Springer

Bonamin, L & Waisse, S. (2014) 'Biology and Sign Theory: Homeopathy Emerging as a Biosemiotic System'. *Journal of Medicine and the Person.* 13, 18-22

Bonamin, L. (2011) 'Biosemiotics and Body Signifier Theory: A way to understand high dilutions'. *International Journal of High Dilution Research*, [S.l.], 10(35) June, 66-76

Cedar, S.H. (2017) 'Homeostasis and Vital Signs: Their role in health and its restoration'. *Nursing Times,* June

Hahnemann, S. (2004) *The Organon of Medicine* (6th ed). B. Jain Publishers

Jorgensen, D.L (1989) *Participant Observation: A Methodology for Human Studies.* Sage

Kayne, S. B. (2006) *Homeopathic Pharmacy: Theory and Practice.* Churchill Livingstone

Kofka, K. (1935) *Principles of Gestalt Psychology.* Lund Humphreys

Phatak, S.R. (1988*) Materia Medica of Homeopathic Medicines.* Foxlee-Vaughan

Ryan, S. (2004) *Vital Practice Stories from the Healing Arts: The Homeopathic and Supervisory Way.* Sea Change

Ryan, S. and Tsamenyi, S.K (2016) *Volta Voices.* Ghana Homeopathy Project, pp. 73-75

Sankaran, R. (2005) *The Substance of Homeopathy*. Homeopathic Medical Publishers

_____ (2007) *Sensation Refined*. Homeopathic Medical Publishers

_____ (2014) *Natural Kingdoms: Healing with homeopathy*. Penguin

Further Reading

Bloom, K., Galanter, M. and Reeve, S. (eds.) (2014) *Embodied Lives: Reflections on the Influence of Suprapto Suryodarmo and Amerta Movement*. Triarchy Press

Goli, F. (ed.) (2016) 'Biosemiotic Medicine: Healing in the World of Meaning', *Studies in Neuroscience, Consciousness and Spirituality*, Springer

Sebeok, T.A. (2001) 'Non Verbal Communication' in P. Cobley (ed.) *Routledge Companion to Semiotics and Linguistics*. Routledge

Whatley S., Garret Brown N. and Alexander K. (eds.) (2015) *Attending to Movement*. Triarchy Press

Wheeler W. (2006) *The Whole Creature: Biosemiotics and the Evolution of Culture*. Lawrence and Wishart

_____

Sheila Ryan is author of the ground-breaking book *Vital Practice* which introduces a mindful and homeopathic supervision to helping and healing professionals. Sheila teaches this approach internationally, these days mainly online. Living in the mountains in Southern Spain with her husband and terriers, Juno and Gloria, Sheila runs *Sea Change Supervisions*. Sheila Ryan B.Sc. Dip. Couns. Dip. Hom. Dip. Sup.is a retired Fellow of the Society of Homeopaths.

info@seachangeuk.com ~ www.seachangeuk.com

The Transformative Body

Using movement awareness for transformation post cancer: a transnational narrative of home

Ditty Dokter

Abstract

As a drama and dance movement therapist, recovering from breast cancer meant reacquiring agency and a sense of ownership of my body personally and professionally. This chapter describes the process of regaining the body, firstly by using weaving to give external form, secondly by rebuilding bodily confidence and thirdly by participation in creative project groups. The aim was to devise and perform autobiographical, site-specific, movement-based work as both a way of exploring and accepting bodily changes and finding a new way of being in the world with myself, others and my environment. Breast cancer's causes are a mix of environmental, social and genetic/family factors, which are different for each individual. These factors echo aspects of Amerta Movement: awareness in its relationship to the body in movement, to others and to the environment. Exploring mortality and a place to die / be at home were part of my migratory movement explorations. As an older migrant this enquiry is an embodied expression by a transnational in the context of 'narratives of home' (Walsh & Nare, 2016).

Introduction

Completing breast cancer treatment in autumn 2011 was the start of a journey of recovery. Treatment consisted of surgery, radiotherapy and medication, all affecting my energy levels, body image and ability to move. Regaining my body was a struggle. Expressing myself creatively through my body was painful and limited; re-acquiring a sense of agency and ownership of my body took time. Experiencing cancer, however supported, is a lonely experience in the struggle for survival (Dokter, Lea-Weston & Thornewood, 2020). A reconnection between inner and outer worlds may be needed. During the treatment I kept a journal in which I drew and wrote about the experience (Wadeson, 2011). The journal became part of my recovery; I wanted to give external form to the experience and needed a creative mode of expression. Prior to and alongside the expressive embodied explorations described in this chapter, I used tapestry weaving and felting to create wallhangings to give form to my experience separate from the body. Examples of these are interwoven with the movement story.

The chapter aims to stay close to the artistic product. As a dance therapist I am interested in dance/movement as a form of data collection (Hervey, 2004), as a dramatherapist autobiographical performance and storytelling are an intrinsic part of my movement vocabulary (Dokter & Gersie, 2016). The enquiry question I formulated related to the form of the creative product and used artistic methods of data collection (movement, images, journal writing). That question was "Can movement interact with two-dimensional weavings in transforming an illness experience?"

Journal

The journal contains poems about arrival and departure in the hospital and cards sent by friends and family. Drawings of the absent breast and the radiotherapy machine are all put together in a type of scrapbook chronicling feelings and experiences of those first six months of diagnosis, treatment and early recovery. I have written and drawn them at varying stages, not chronologically, but when they arise and when I have enough energy, I give them form. Associated childhood memories of hospitalisation are also translated into words and images; the past juxtaposed with the present. I remember the toys I received during my hospital stay in childhood, now replaced by many cards and 'get well' messages. These give the message: you are not alone, you have companions on this journey. For all that, I ultimately feel that I am alone. I am stuck in bed again, not able to move, in an isolation room. I am ill…again, I am stuck in bed…again, I am alone….again. In this chapter excerpts of the journal are quoted or paraphrased in *italics*.

Fig.1: *Cancer Journey.* Wall hanging, felted images on post-surgery 'blanket'

Weaving

I was very aware of the connections between my childhood and adult illness experiences, a sense of revisiting trauma. As an adult, I wanted to find ways of helping myself to both accept and move on from the powerlessness of those experiences. I needed a creative mode of expression. Using my body immediately post treatment was fraught with difficulty, so I looked for an alternative medium for exploration at that stage. I was grateful for the help of a friend in reacquiring textile arts skills that I had used in an earlier period of my life. The resulting wall hanging created by the first anniversary of completing treatment (Figure 1) was an important step in starting to feel stronger.

Movement

I continued tapestry weaving, but also felt the need to regain my expressive embodied self. I have worked as a drama and dance movement psychotherapist for 35 years and embodiment is an integral aspect of my practice. Moving was part of my self-help strategy to survive what often felt like an invasive and harsh treatment of the cancer; my own cells attacking my body. Alongside hospital treatment my local Maggie's centre offered yoga and Qigong and these had strengthened my functioning body, but I was unsure whether I could regain a more expressive embodiment now that I needed to walk with sticks and to move gingerly to ameliorate the pain in my joints.

I had in the past been interested in Sandra Reeve's environmental movement workshops. A year after completing treatment I retired from my NHS employment on health grounds and my partner offered to drive, enabling me to attend some initial Move into Life weekend workshops. Awareness through movement in Move into Life/Amerta Movement was closer to a form of movement meditation than my previous practices, exploring a more meandering path where movement awareness led to discovering connections within myself, between myself and others, between myself and the environment.

Sandra facilitated me to adapt my movement to what was feasible, as I tried to accept my more fragile bodily self. Wanting to extend the moving indoors to working outdoors in relation to the environment, I took part in a week's workshop on Autobiographical Movement in the Burren, Co. Clare, Ireland.

Crystallisation 1 – Workshop Sharing

> *Two of us move together in the cave: we alternate moving singly, our stories alternate but are intertwined. I use my sticks to get across the boulders on the beach to the cave. I tap with my stick against the cave ceiling, the stones around me; can I move, can I climb, am I stuck here, can I get out? Slowly, at times crawling, at times using my sticks, I manage to traverse the obstacles to the big flat boulder at the entrance of the cave. I heave myself up to standing position, I can stand without my sticks. I can climb down without my sticks and get to the sea. Singing, moving, I can move!*

Moving as part of a group, with one other group member also moving with difficulty in the challenging terrain of limestone pavements and rocky shores, resulted in a crystallised movement improvisation in a sea cave around the theme of accepting limitations. Sticks remained objects of ambivalence, both creative body extensions and impediments which I was tempted to throw away – a creative theme to hold the paradox whilst a resolution may be found over time. Being able to interact with the environment as non-judgemental witness, as well as with other movers as supportive companions, was an important component of my recovery. Having completed that challenge, I felt more able to engage again in expressive movement exploration related to autobiographical material, two years after the end of my cancer treatment.

Movement experience: awareness through movement

In autumn 2014 I joined the one-year creative project group, wanting to continue to work with the theme of 'accepting limitations'. My question at the start of this project was whether movement could interact with two-dimensional weavings in transforming an illness experience. I had been weaving landscapes that had been important to me throughout my life in the Netherlands and the UK: the forest, sky and seascapes of my two home countries (www.whitwellweaving.co.uk).

> *I push away from the wall, try to move bending backwards whilst following the beams of the ceiling, pushing my flexibility. In this pushing for flexibility I become aware of my stiffness, I do not want to yield to it. I want to keep striving for the flexibility I used to have. I meet M. who wants me to play with her, I resist, try to stay with the stiffness. Sandra starts to play the drums, I try to run, move fast ... I*

> *cannot run ... I stop, slowly I move again, find myself speeding up my walking rhythm.... I cannot run, but I can speed walk. I start to walk round the hall, look round the space, some group members have started to play with chairs. I sit on one of the chairs, the relief of sitting, arriving, staying with, rather than striving. I sit for a while, in stillness I become aware again of the others and want to interact. I can play peek-a-boo, put the chair on my head, my arms stick awkwardly though the stretchers connecting the chair legs. I move stiffly, but I move; I am trapped in the chair cage, but I can play.*

Moving with others and objects in the indoor space as part of the movement exploration became associated with the movement dynamic of weaving: pushing, pulling, reaching, taking, holding and placing. I was interested that 'reaching' for me became associated with striving, in contrast to my theme of accepting; whilst 'holding' needed permission to keep: in movement I tended to want to let go again. Putting and placing led to installation, which created more interaction with others. The interaction with others became more important for my exploration than the interaction with the tapestries, so I started to explore nests, where each group member could have their own nest/habitat and move in and out of that space. I became aware of migrant identity, shared with others in the group. I found that in trying to accept, I tended to strive to move beyond limitations. Balancing the joy of 'I can do that again' with being kinder to my body, not pushing beyond my limits.

> *I move outside in an ancient place on the brow of a hill; there is bracken, high grass and a small copse. I need to be mindful of my legs' ability to carry me. I need to carefully place my feet and use my sticks. However, I can tap on rocks and trees evoking new sounds, my sticks can be body extensions, I can stretch, overbalance and feel safe. Leaning on my sticks I start to look around, noticing the distant sea. With my sticks the world can be bigger. I start to move around, meet others. I am offered and accept help to traverse a very muddy patch with lots of awkward tree roots, accessing a new field where others are sitting, watching and moving. I can join them.*

Moving in an outdoor environment I found it easier to accept (my) limitations, was forced to take care of them on uneven ground; my movements needed to be small and focus on stillness and moving on different levels, everyday movement.

Body and Awareness

I needed to be mindful of my legs' ability to carry me, the placing of my feet, the use of my sticks as body extensions for exploration, not just as beasts of burden. Sticks could be played with, become props, characters or musical instruments. Using my sticks, being supported, I noticed the open vistas in contrast to my smaller movement world without them. In an outdoor environment I also needed to accept support from others to traverse certain terrains.

Travelling in different terrains broadened my movement vocabulary, gave me a chance to widen movement patterns, to move from more linear directional movement to a freer exploration of space, just as the support of the sticks opened up new vistas.

> *I have hung the weaving on the back of a chair and am moving in the hall across the parquet floor. I crawl, roll and lie still looking at the different squares of the wood, following the edges of the squares with my hand, my elbow, my knee, other body parts. I come across a patch that is warm and lighter than the others, I move in it, lie still in it, start to roll from left to right with my eyes closed, feeling the warmth. I come up on all fours and start to crawl keeping to the light patch, stopping when it disappears. I open my eyes and come to standing, travelling, moving in the patch of light, my arms and body following the long diagonal rays of the sun coming in through the window.*

Emotionally I was processing my own illness experiences, as well as my father's death, after a progressive terminal illness. Through movement, I found that being in the here and now also connected me to past memories. Indoors again I moved with a tapestry of a forest path, a route I had walked with my father many times to visit the 'ven'. The movement was slow and mournful, interspersed with childhood memories of skipping and being held by the hand, mine going up, his down, when walking together through the dark old forest toward the distant light reflected on the water (Figure 3). Moving in and towards the sunlight through the window in the hall translated my childhood experience to the here and now movement experience. Past and present interwove when the memory was given space through embodiment.

In the tradition of Amerta movement Sandra asked us to explore moving from a sense of the sacred, which for me connected to spirituality. I found myself belly down dragging myself across the floor singing a one-line chant, feeling a strong sense of loss and longing. The movement and vocalisation were repetitive, a re-enactment of feeling lost and alone, but also searching, calling.

Fig. 2: *Veluwe Forest Path to the 'ven'*. Tapestry woven from fleece and wool

Awareness of my mortality and wondering in which of my two countries I wanted to die, I became aware of the need to find a spiritual tradition which could facilitate that rite of passage. My ambivalence in the process of being accepted/accepting was embodied and expressed through songs to accompany that embodiment. The songs were in Dutch and French from various times in my life, exploring in their lyrics differences in language, landscape and spirituality.

Body and Awareness

Crystallisation 2 – Invited Audience

I move outdoors in a wild garden offering some protection.

I accompany myself with certain songs from my life, starting with children's songs.

The weavings are strung between the trees, those with the Dutch landscapes on one side, the UK landscapes on the other.

Fig. 3: *Finding Home*

The movement is unaided, without my sticks and is chronological; starting from childhood, to migration at young adulthood, my encounter with refugees, receiving and finding a place for nests and a final section relates to how the illness experience raised new questions about 'belonging' (both spiritually and geographically). The crystallisation is called 'Finding Home' and is performed in front of an invited audience and includes some audience participation. The group members at a given moment become 'refugees' having provided birds' nests from their places of home (which birds had vacated) – those nests are connected through a three-dimensional woven web created by moving between the nests during the crystallisation (Figure 3)

Movement experience continued: Movement Art

In 2016/17 I took part in the second project group. The Brexit vote had renewed my question of belonging. My explicit aim was to create a public performance round this theme called 'Come or Go'. This developed the movement practice and crystallisation from the first project group. The nests from the first performance were replaced with a woven boat (Figure 4). I decided to use the beach as the performance site. The constant flux between land and sea and the constantly shifting landscape of the beach reflected my questions about belonging and welcome. Although my ability to walk had improved, the surface of pebbles made walking hazardous and I needed two helpers to facilitate my entering and exiting the sea during the performance. This reflected my own ambivalence: did I have steady ground under my feet in the face of Brexit, was I being assisted to leave or to remain? The performance took place on a public beach. Erosion means the cliffs are crumbling progressively, as temporary as the division between land and sea. In contrast to my contained first project crystallisation, this performance was witnessed by strangers as well as invited audience members. I chose to perform at the turn of the tide where the sea starts to return and the land disappears. The performance took an hour and the early writing in the sand had disappeared by the end.

Fig. 4: *Come or Go*

Crystallisation 3 – Public Performance

The promenade performance starts with me drawing in the sand COME on one side of the beach, GO on the other, an outlet pipe separating the two. I walk over the pipe with my sticks towards the audience, inviting them to walk the words as a path with me, first come, then go. At Go the boat awaits and I invite the audience to join me in launching the boat in the river which flows across the beach. The boat is obviously not seaworthy, but the launch is tried in several places, culminating in me swimming it out to sea. It returns, I rejoin the audience apologetically on the beach, murmuring "I could not go, I tried...". In the three places of the promenade performance I sing verses from Jules de Corte: "Ik zou wel eens willen weten = I really would like to know", culminating in my final arrival back on the beach when I sing:

I really would like to know	*I really would like to know*
why the cliffs are so high	*Why the cliffs are so high*
maybe to gather the snow	*Do they really support the sky?*
to protect the valleys from cold	*When do they live or die*
or maybe they are a support	*When I live or die*
for the arc of the sky	*Do I want to know?*
Maybe that's why the cliffs are so	*Maybe I really don't want to know*
high (repeated)	(repeated)

Movement awareness and transformation

My question at the start of these projects was whether movement could interact with two-dimensional weavings in transforming an illness experience. As creating a tapestry had been the first step in processing the illness trauma, it was important for me to incorporate the tapestries in the transformation from illness to healing. They were physically present in both crystallisations; strung across the garden in the first, as laminated photos in the second, with the originals exhibited in the village hall nearby.

The movement awareness enabled me to become aware of my striving and pushing, so it became possible to work more on acceptance, being in the moment with what is. The isolation of the illness experience was ameliorated by moving with others, the movement awareness of needing aids, in the form of sticks and companions could be played with and given form in the crystallisations. My world could expand again by moving in outdoor environments and I could regain a sense of agency in companionship.

The paradox was that in doing so and in gaining a greater sense of finding home, even an acceptance that this country was the place where I could live out my days, suddenly external political circumstances put the question of 'where is home?' in a different perspective.

When the issue of separation arises, transitional objects can enable attachment transition; for me these were the wall hangings that gave external form to both the illness and to migrant journeys. The tapestries enabled me to connect the different landscapes of my life. The emergence of three-dimensional woven nests in 'Finding Home' and the boat in 'Come and Go' were also crucial transformative props. One of the project group members asked if the 'softness' of the tapestries could be connected to the 'harshness' of the migration experience. The 'Finding Home' nests were a mix of soft moss and tougher twigs; the 'Come or Go' boat, woven of willow withies and branches, was a harder shell, but permeable. Its lack of seaworthiness became a poignant reminder of the lack of seaworthy vessels carrying refugees, connecting the perilousness of life and death to both illness and migration.

Movement made me aware of, and gave embodied form to, the impact of my fragility, both as a childhood and as an adult experience. Moving in the here and now helped me to transform some of the childhood experience of helplessness, the inability to move. As an adult, my awareness of mortality evoked the question in which of my two countries I might want to die. I found myself using both my first and second languages to give voice to the silent movement experience. As my illness was very much an embodied process, movement needed both a voice and a product to hold the experience.

The songs in the first crystallisation were sung only in Dutch, with French to conclude (translations in the programme). This was meant to create distance between performer and audience and to represent the experience of being a migrant, as did the audience participation, when they were told to move from one place to another, questioning the safety of place.

In the second performance I sang first in Dutch, then in English translation. The transition between the two was from private crystallisation to public performance, as the second performance was in a public space and intended to portray my personal experience in relation to public political issues.

My private illness experience was translated into the more public migration one. The fact that the people in the UK landscapes I love tended to vote leave, in favour of Brexit, emphasised my feelings of alienation, not being welcome and questioning my wish to stay in the face of that feeling. Margit Galanter (in Bloom *et al.,* 2014) said that her movement exploration of 'leaving' revealed for her the paradox that leaving can keep you stuck in a process, even though you want to run away. Galanter's emphasis on a felt sense of embodied context meant for me that my moves to leave kept me stuck. My choice to stay gave me freedom to move.

The first crystallisation was intended to be a way of sharing experience in order to transform that experience, but I identify with Lise Lavelle (ibid) when she says that her performance made her learn new skills and discover new ground, personally and professionally. Dramatherapist Robert Landy (1993) writes that a role needs a story in order to find expression, whilst Gersie (1991) emphasises that stories provide a process to give form to our life and voice to our inner experience; to shed light on the contrast between inner and outer worlds. Through the movement exploration I found I needed to use my voice. The songs gave me the means and the confidence to 'voice' my movement. The embodied performance of my own story to transform the shame of illness/disability, exile and trauma (Rubin, 2016) was empowering.

Finally, the importance of the group experience, as support, witness and partners in improvisation, as part of a reciprocal collaboration, (as opposed to working on my own) enabled me to struggle with accepting support, part of my acceptance of dis-ability. It also enabled the co-construction and recon-struction of memories, to share across differences in culture (Dokter & Gersie, 2016). I could be welcome, whilst questioning the nature of that welcome.

The beloved outdoor landscapes in my adoptive country supported the re-finding and transformation of my expressive movement. Working with others enabled the process of acceptance. Both facilitated paradoxically a strong sense of belonging, in circumstances where isolation and alienation were paramount.

Conclusion

The initial artistic inquiry question was "Can movement interact with two-dimensional weavings in transforming an illness experience?" The weavings were both a beginning of externalisation and a bridge into embodiment. Movement exploration and performance enabled a means of transforming the illness and isolation/alienation experience. The weavings provided a bridge between the functional and expressive embodiment practices and are therefore a crucial component of the artistic inquiry both at the onset and in the culminating performances.

Acknowledgements

I would like to thank the members of the Move into Life Project Groups 2014-2015 and 2016-2017, Sandra Reeve and David Tatem for enabling the work described here. I also thank my Anglia Ruskin University colleagues Mandy Carr, Julianne Williams-Mullen, Sasha Nemeckova and Naz Yeni for co-creating the Nomadians theatre company and performing the parallel 'group crystallisation' 'Welcome! Question Mark?' (Edinburgh Fringe, August 2017).

References

Bloom, K., Galanter, M. and Reeve, S. (eds.) (2014) *Embodied Lives: Reflections on the Influence of Suprapto Suryodarmo and Amerta Movement.* Triarchy Press

Cochrane review (2017) Can music interventions benefit cancer patients? www.cochrane.org

Dokter, D. & Gersie, A. (2016) 'A retrospective review of autobiographical performance in UK dramatherapy training' in S. Pendzik, R. Emunah and D. Johnson (eds.) *Self in Performance.* Palgrave Macmillan

Dokter, D., Lea-Weston, L. and Thornewood, T. (2020) 'Secondary Traumatisation and Therapist Illness' in A. Chesner & S. Lykou (eds.) *Arts Therapies and Trauma.* Routledge

Gersie, A. (1991) *Storymaking in Bereavement.* Jessica Kingsley

Hervey, L.W. (2004) 'Artistic Inquiry in Dance Movement Therapy' in R. Cruz & C. Berroll (eds.) *Dance Movement Therapists in Action.* Charles C. Thomas

Landy, R. (1993) *Persona and Performance.* Jessica Kingsley

Reeve, S. (2013) *Body and Performance.* Triarchy Press

Rubin, S. (2016) 'Embodied Life Stories: Transforming shame through self revelatory performance' in S. Pendzik, R. Emunah and D. Johnson (eds.) *Self in Performance.* Palgrave Macmillan

Wadeson, H.C. (2011) *Journalling Cancer in Words and Images: Caught in the Clutch of the Crab.* Charles C. Thomas

Walsh, K. & Nare, L. (2016) *Transnational Migration and Home in Older Age.* Routledge

Wolford, L. & Schechner, R. (1997) *The Grotowski Sourcebook.* Routledge

---

Ditty Dokter PhD is a drama, dance and group analytic psychotherapist. After a working life in the NHS and universities, she is now 50% self-employed (therapy/supervision/research/training) and weaves, writes and moves in the other 50%. Ditty has published various articles, chapters and edited books. A companion chapter to this one is (Dokter *et al.*, 2020) referenced above.

dittydokter@googlemail.com

The Unveiled Body

A Release from Obsessive Compulsive Disorder through Dance Movement Psychotherapy

Céline Butté

Abstract

Like a piece of furniture covered up and forgotten in an attic corner, our own body may at times feel quite familiar yet hidden from us. In this chapter, vignettes from an interview with Emuru about her dance movement psychotherapy journey illustrate how attending to the body's way, to the story it holds and yearns to express, and to the client's constructed verbal narratives, opens up spaces for deep and long-lasting healing from significant emotional, physical and psychological difficulties, in this case Obsessive Compulsive Disorder. This is a healing that includes our somatic, psychic and relational existence, and which Emuru called an *unveiling*. Awareness and the body are at the centre of this single case study which offers insights into an approach to psychotherapy underpinned by core principles from Contact Improvisation, Body-Mind Centering and Sensorimotor Psychotherapy.

Fig 1: *Unveiled*

C. *If you were to give a title to your image what would it be?*

E. *Probably 'unveiled'. This is like a veil that I was stuck behind; I didn't even know I was behind it … and I thought that I understood everything but actually I needed to move it away; and it kind of opened up both body and mind I would say.*

This writing is inspired by an interview with Emuru, a young woman who, when she started dance movement psychotherapy, was yearning for relief from relentless internal demands that manifested in rigid and repetitive ways of completing daily tasks. Brushing her teeth and getting out of the house had become dreaded moments when she would get caught in a battle of will with her own self. Obsessive Compulsive Disorder had taken control of Emuru's life and had led to mental, physical and emotional exhaustion.

This chapter is organised around three distinct and interrelated domains that Emuru and I gravitated towards in our work together: *meeting the ground, carving new paths* and *finding the words.*

Meeting the ground

Week after week, Emuru walks into the therapy room, releases a deep sigh and reaches for the ground.[1]

Arriving

As she yields into the ground and releases her breath, Emuru's body is telling us that she needs to let go and be received. Attentive to the weight of her sigh, I encourage Emuru to let her body guide her. Our work has begun. *'She lies there for a moment, gently brings her legs to her chest, releases her breath again, pauses, rolls slightly to one side then to the other, stretches an arm …',* letting her body lead the way and letting herself feel and sense the aches and pains and the places of ease. This takes all of her attention and transports her into uncharted territories where, as she explains in our dialogues later, she finds unexpected relief and renewed vitality.

In the safety of the therapeutic relationship, Emuru surrenders to gravity and to herself; I guide her to feel her body in contact with the ground, her own edges and the life within. My intention is to invite a slowing down and shift of attention from her mind to her body, whereby she tunes in with what she feels she truly needs, rather than following a cognitively driven belief of what is right.

Through this work, we activate Emuru's parasympathetic nervous system. Our parasympathetic nervous system (PNS) is the part of our nervous system that we need for recuperation. It works concomitantly with our sympathetic nervous system (SNS), the part of our nervous system that we need to get into action. The PNS and SNS depend on each other and make a whole.

As she yields to her body and tracks what needs to happen next, Emuru deeply feels the importance of balancing the activation of both the PNS and SNS. Gradually, she learns to appreciate the quietly busy place of her PNS and how to navigate life, tapping into the wisdom of this dance between her SNS and PNS.

In daily life, being up and active is given much value, whilst resting may be seen as a waste of time, something less important than activity; we may even feel guilty when we do take the time. So we keep going, taking in

---

[1] Sessions welcome the use of the ground. Often clients are invited to take their shoes off, make themselves comfortable, use the ground or sit on chairs, as they prefer.

more, without letting go of enough. This creates a tension, which our body expresses through becoming more rigid. Over time, over many years, we build habits of holding that impact the whole of our life; to the point where our body shows us, in its exhaustion or through pain, that this is not viable any longer.

Although the sigh and reach for the ground remain the start of our sessions for many weeks, the initial urgency to release fades away, to be replaced by a sense of easing into the support that the ground provides. Emuru also begins to integrate this quality of presence into her life beyond therapy. Increasingly comfortable with giving attention to her needs to land in her body and follow its way, mindful embodied processes become as valued as times for planned activities. My guidance is an external mindful presence which in time awakens Emuru's own mindful attention.

Meeting the self, listening to the body

> E. *I mean just listening to ... for example right now I would like to just lie down... and listening to that. And ... I mean listening to the body in a way; like 'this is what it needs right now' and much more so than before. And also the ... bringing it back to the body rather than just left in the head. If there's a situation; trying to experience it with the whole body, being aware of it, what effect it has in other places rather than just the mind.*

Attending to the sensations in the body and attending to our physical and physiological needs is essential in making self-nurturing decisions. During many sessions I asked Emuru: *What do you feel in your body? What part of your body feels comfortable / uncomfortable? What needs to happen to make the discomfort a little less uncomfortable? How do you know you are feeling better?* Thus we open up to the dance of putting sensing-and-moving first. As Miranda Tufnell explains:

> Moving from sensation (...) awakens a quality of dreaming in which the body softens and unfolds and the imagination begins to flow with images, memories and stories. Through their poetic resonance, these images make visible forgotten and often neglected aspects of who we are. What emerges has the power to open up new perspectives in how we see ourselves and thus affect health and vitality.
>
> (Tufnell, 2010: 33)

This embodied-sensory realm of expression is a 'potential space', a play space where the life beneath our skin is prioritised. This does not mean that we are not allowed to feel or think, it simply means that sensing and moving are enrolled alongside our feelings and thoughts, inviting more fluidity and connectedness between our physical, emotional and psychological experiences.

In an action-oriented and mind-driven world, we easily slip into placing our physical needs last. We may even have become so alienated from our bodies that we are unable to recognise what they can tell us. In order to know how to look after the self, we have to re-learn something we knew instinctively when we were little: our body's unique way of telling us what it needs and wants. Therapy sessions were opportunities for Emuru to practise being present, through paying attention to her body, and thereby understanding her needs and wants and taking action accordingly.

This work nurtured the possibility of Emuru listening to her inner self and being curious about who she is. During these times, I guided her to feel her weight, be supported by the ground, so she could stop overthinking and let herself be. At times, Emuru would recognise blocks and challenges through this embodied inquiry and would process the emotional charge attached to them.

Contact Improvisation: a source of inspiration

The meeting-the-ground I present above, through *arriving, meeting the self and listening to the body,* is inspired by Contact Improvisation (CI). CI is a dance form that I have been practising for many years; it offers a powerful experience of landing in the moment, with my choices, and in relation to others. CI, as its name tells us, includes contact, more specifically physical contact. The physical part of CI, the use of touch through leaning into another, receiving another, pushing, pulling, brushing past and being touched by others, opens the door to the relational field in a concrete and tangible way. It is a context where I have made numerous discoveries about my own capacity and limitations, and where I have played with new possible ways of being and relating. In an early film about CI, Steve Paxton, one of the founders of the form, mentions *gravity, the centrifugal force, support and dependency* as the basis of CI (1987).

Although I may not directly use CI in my psychotherapy work, it is a tool I draw on in relation to my clients. Psychotherapy, to me, feels so alive and makes complete sense when seen through the lenses of an embodied practice such as CI. The experience of *gravity* relates to this landing in the body I have just mentioned; the *centrifugal force* is what I would describe as 'going with the flow' (in the sense of following a natural force), letting go and letting be; *support* is the leaning into a concrete support within our environment, or another person, literally and metaphorically; it is about trust. *Dependency* is a reminder of our interdependency as human beings and our inevitable dependency on internal and external factors to function well. Of course, CI does not translate into psychotherapy, yet CI calls our attention to fundamental aspects of our shared human experience that serve my psychotherapeutic practice.

Sensing our way into knowing

When attending to meeting-the-ground during the psychotherapeutic encounter, instead of focusing on our thoughts and keeping ourselves busy with many activities, we practise giving our physical being credit for what it is drawing our attention to. There is a refreshing and enlightening simplicity to the process. Through our movement we either *go towards* or *move away from* something; fundamental responses of our existence that represent what we say *yes* and *no* to.

When we enter this world, our survival mechanism drives us (i.e. we are hungry: we cry, and if we are lucky enough, we get fed), it is an instinctual knowing, a pre-conscious experience that becomes part of our unconscious as we mature. As our cognition grows, a wonderful and sometimes confusing array of alternatives also emerges; i.e. '*I am hungry, let me think: do I have the time? should I finish this task?*' Through embodied processes of inquiry into what we go towards and move away from, we have a tangible experience of what we say *yes* or *no* to. Keeping our attention with our body, our sensations and movement, we revisit the choices we make and reconnect with the parts of ourselves we have long neglected. This awakens the possibility of making new choices and of living life with renewed awareness, in a way that serves us better.

Initially, for Emuru, landing and grounding occurred in her experiencing herself fully during the therapeutic encounters: sensing into her body, following its way, meeting blocks and places of holding and processing the emotional charge attached to these. Gradually, she infused other areas of her life with this *in the moment compassionate inquiry and self-nurture* which revealed to her what she deeply wanted and needed to say *yes* or *no* to.

Carving new paths

We live life in patterns established by internal and external forces that we respond to. The way the world meets us and the way we meet the world is a dance between our inner and outer realities. Each of us has our idiosyncratic ways of doing things. Of course, our days are not pre-set and these patterns can change. However the way we live life is embedded within a complex web of experience, memory and choices that imbues an element of predictability and repetition in each of us.

Someone with OCD has become driven by patterns and ways of engaging with life that no longer serve them. In the work with Emuru, we demystified the question of habits and inquired into the many rituals that were driving her life. We wondered: *What is this serving? Is this useful? When did this start?* We sought to discern between what were healthy and unhealthy rituals.

Soon, we elucidated that Emuru's daily rituals were particularly active when Emuru felt tired or emotionally overwhelmed and had developed in an attempt to feel a sense of control. Below, I give examples of three aspects of the embodied process of changing habits of daily life which addressed these physical, emotional and psychological issues of tiredness, emotional overwhelm and control.

Repatterning

> Unlearning from the nervous system. Relearning from the cells.
> (Bonnie Bainbridge Cohen)

When she started her therapy, Emuru had already crafted a promising career for herself in the family tradition of Kintsukuroi[2], an art that she practises with dedication: she spends hours bent forward in her workshop, shoulders tucked in, hands focused on her craft. She also shows her work at fairs and exhibitions, where she stands for hours at a time. During numerous sessions, we worked directly with those postural habits, paying deep attention to subtle and louder cues held in her physical being, without interpretation. For example, we explored what was going on in her joints, her sense of feeling restricted or limited, and her needs for a stretch, repressed or met.

---

[2] Kintsukuroi is the Japanese art of repairing broken pottery with lacquer dusted or mixed with powdered gold, silver or platinum

The aim was to address Emuru's physical exhaustion directly and offer practical solutions to physical pain. In those moments, we would discuss her working day and explore postural, movement and breathing changes that she could implement at her workshop in order to be more comfortable, however busy she was.

Our work with patterns and self-care also included the acknowledgment of the different seasons in Emuru's life. We considered how she could pre-empt the intense and the quieter periods of work, the creative times in her workshop and times at her desk, as well as her needs and mood when pre-menstrual, so that she could flow more consciously between these different and equally valuable seasons. A growing awareness of this cyclical bigger picture enabled Emuru to anticipate those seasons and give herself permission to ease into taking moments of much needed rest.

As we had identified that tiredness fed Emuru's OCD, this solution-focused part of the work was crucial in breaking a pattern of driving herself to exhaustion through self-sacrificing devotion to her work. These were times of inquiry into what it means to be present that shook up old habits and nurtured new self-caring pathways.

Resourcing

To attend to what challenges us most, we need to have a safe base within ourselves, *a happy place*. In order to build one, I invited Emuru to remember events from her life when she would have felt relaxed and at ease. When we found a particularly vibrant memory, I took her back to this place of nourishment and satisfaction through visualisation, inviting her to bring to life, in the here and now, the felt sense of this memory.

The literature on trauma and the body which I am familiar with through Sensorimotor Psychotherapy speaks at length about the importance of homing in on such significant memories (Fischer, 2019), calling them our *resources*. Connecting with a resource when faced with a challenge – from a recent event or old memory – is like an anchor. It offers a safe position from which to apprehend emotional overwhelms. For Emuru, floating on the water became one such place of contentment so, to build this internal resource, I invited her to linger in the joyful feelings of floating on the water.

A felt sense of trust that melts a superstitious belief

Early on in her therapy, Emuru recognised that part of her OCD revolved around the superstitious belief that *if she did not do this or that x amount of times, something terrible would happen to someone she loves.* The significance of the resource of floating on the water became clearer to Emuru as she described the possibility of being both in and out of control when lying on her back on the water: *the water is an unpredictable external force, and Emuru is at the mercy of the timing of the next wave. Yet, in those moments, she feels safe relaxing on the water, as her body knows how to respond at any given moment to the movement of the element beneath her.*

Floating on the water is a cherished familiar place where Emuru unconsciously plays with the felt experience of remaining quietly alert and responsive so as not to be submerged by the water; a knowing through the body of what it is like to let go of control without collapsing. Bringing into awareness this concrete felt memory, stored in her body, of what it is like to feel safe even when she cannot control everything, demonstrated to Emuru that she knows how to trust herself. A now accessible resource, this memory became the springboard for her to trust events that she could not control. This experience is akin to the *centrifugal force* of CI mentioned earlier: this idea of letting a natural force show us the way so that we let go and let be yet stay present and attentive. Floating on the water is an embodied knowing that became a reference, a go-to place for Emuru. When something she could not control was about to happen to those she loves (for example, if they were about to fly somewhere), she could ease into the somatic sense of this floating-on-the-water memory instead of whirling into debilitating thoughts that fuelled anxiety which in turn triggered obsessive patterns of behaviour. Bringing this memory into her felt experience and consciousness loosened the grip of a powerful superstitious belief that was at the core of her OCD.

Embedding new habits into life

> *I feel like it's been very practical, it's been proactive. It's like 'why are you just talking about 'floating makes me happy'? just 'do something about it'. (...) And it's so important to be like 'so why don't you just sit down for 5 minutes and do some research now?'....*

To consolidate the experience of floating on the water, I invited Emuru to think about ways of bringing the concrete experience of her enjoyment of water more fully into life. This is what she calls the proactive part of her therapy. She researched local swimming pools, something she had wanted to do since arriving in London a few years before, yet had not done. This link into the reality of Emuru's life supported her to commit to herself. In those instances, I was holding the role of an encouraging other, who was taking what Emuru likes seriously. Over time Emuru awoke this capacity in herself. I feel that as therapist it is important that I sometimes facilitate such self-caring homework as it fosters a shared attention to something important that needs to be invested in with more discipline and commitment.

In this aspect of the therapy we engaged both Emuru's neurological (internal) world, and her relationship with the outside world. As she developed new patterns that attended to posture, movement and breath and connected with resources, Emuru freed up new, less tiring ways to get through the day. The felt sense of safety embedded in her body and consciousness through the resource of floating on water provided a more stable ground from which to meet the unpredictable. Thus somatically driven mindful processes that took Emuru's needs and preferences seriously started to replace habits of physical holding and relentless controlling cognitive patterns: a more organic ground from which new and more nurturing life habits could emerge.

Finding the words

> Language is a skin: I rub my language against the other. It is as if I have words in place of my fingers, or fingers at the tip of my words.
>
> (Barthes, 1977)[3]

The words we choose contain our story, they impact both ourselves and others. Inadvertently we may get used to telling our story in a particular way, stuck in a 'then experience' of an event rather than open to the life a memory holds for us at any given moment. In this section, we consider how attending to *what* is said and *how it* is said, has the potential to free us up from the shackles of habitual ways of expressing ourselves verbally.

All is not what it seems

> *Behind this veil ... before ... it felt like a lot of chaos in my mind and a bit like closed. And just not knowing what to do with my emotions. (...) At that point it felt that there was only one problem: my OCD;*

---

[3] *Le langage est une peau: je frotte mon langage contre l'autre. C'est comme si j'avais des mots en guise de doigts, ou des doigts au bout de mes mots.*

and that if that was fixed with medicine then I'd be done. But it's so much more than that and that's why I feel like that veil is so good an image because It felt like a heavy wall actually, first; and I just didn't see it was something I could pull. and then once I'd figured out it just was like silk, it was like 'WHAT!' ... It had a transformation.

The imagery of the wall that she could not even see may be understood as the times when Emuru was not aware of what was going on for her internally. As she started her therapy, it felt as if Emuru was an uncomfortable and confused spectator to the unfoldings of her life – inner and outer. She had got stuck and identified with a story about who she was, both in the way she was going through her days and in how she spoke – or avoided speaking – about her experience. In her drawing, we can see that the start of Emuru's therapy enabled her to see that she was stuck behind a wall. This beginning marked her arriving in a space-relationship within which she could feel safe to speak her experience of stuckness, despair and exhaustion and let those words hit the air[4]. A naming of 'what is' that created an immediate relief: the wall transformed into a veil.

Speaking as an anchor for change

Considering her drawing (Figure 1) further, Emuru describes the image in the left corner as her drowning in her tears. She continues: *What could be added here definitely is me floating on the water as a contrast,* and on this note she draws herself floating on the water on the right-hand side of the picture. At this very moment in her interview, Emuru anchors the deep work done through *resourcing* and *embedding new habits into life* (as described above), into the present. Through the medium of drawing she creates and articulates a new narrative. Embedded in her psyche now, along with what was once an unconscious, isolated and overwhelming image of drowning in tears, is the peaceful, contented image of floating and resting in familiar loved waters: two poles that hold a spectrum of potentials through which Emuru can meet her suffering with a sense of trust and safety within herself.

As changes take place at a somatic, physical and emotional level, our cognition can be the last to catch up. Allowing a new self-caring and grounded way to see and speak to an experience is a deeply activating process that consolidates therapy work, grounds it in the present and helps our brain update its records.

---

[4] As self-help author Karen Salmansohn would put it.

Allowing the story to change

> E. (...) But the important thing is that doesn't mean life is just sunny and vacation and ladida. It was like there's still lots of things happening and tears being fallen and thoughts.

I believe that my role as DMP is towards raising awareness of how we seek nourishment and comfort – from within or from outside of our self and become curious about what we say *yes* and *no* to. Although 'balancing' is an aesthetically seductive term here, it is not adequate to describe the kind of journey DMP offered Emuru: to avoid a fake sense of transcendence and an intellectualised view, we must experience the ordinary, the dissatisfaction and the wobbles of daily life; we must feel and sense into these so as to allow a shift to emerge. Mindful attention to our body and to movement is a doorway into such a connection with life in its full imperfect sense. This, I think, is what Emuru talks to in the excerpt above: the unveiling, however revelatory, was only the beginning of a new way forward. In returning to the therapy week after week, to speak to, feel and sense her experience, she gave herself a chance safely to practise delving into this deeper, more alive way of knowing.

As mentioned earlier, caught in her thoughts, Emuru was a detached spectator to her own life. Helpless, she was at times wordless and felt pulled by internal and external forces that she did not know how to negotiate. Sessions became a space to slow down and explore new ways of talking about what was going on. Focusing on contextualised details, with little interest in generalisations, however seductive these may be, Emuru began to feel into the subjectivity and time specificity of her experience. As she allowed a dynamic, committed and soft kind of holistic attention to unfold, rigid ways of being and fixed storylines dissolved. Emuru started to let herself resonate with what felt more alive and true to herself at any given moment, which in turn nurtured a fertile ground from which to communicate with others, in the moment-to-moment experience of any relationship.

In our work, Emuru returned, or maybe even arrived to herself. Over time, her curiosity about her own movement in relation to her environment and to others grew and she developed a somatic awareness practice or *movement reading* as Sandra Reeve calls it (Bloom *et al.*, 2014: 68) that was embedded in her daily life beyond therapy. Lingering in the body nurtured a flexibility of attitude that expanded her ability to tolerate paradox and difference, as well as the known and the unknown. She was learning to live in the moment and to trust her capacity to live and tell her story rather than dismissing her experience and identifying with fixed narratives.

Conclusion

In this chapter I have shared how attending to the body's way, to the story it holds and yearns to express, and to the client's constructed verbal narratives, opens up spaces for deep and long-lasting healing. A healing that includes our somatic, psychic and relational existence. For Emuru this healing melted away the powerful hold that certain rituals had on her.

As she meets the ground and is received by it, Emuru and I arrive at her therapy. Then, quite early on, she recognises that all is not what it seems. Through what she calls an unveiling, Emuru's perspective shifts and her awareness widens; she continues to land in her experience and finds a safer ground within. Emuru's inner witness, a new companion for life, awakens within her through my witnessing and acknowledging of her without judgement. Together we navigate the waters of her emotions, staying with the body more and with thoughts and interpretations less. As we recognise that her professional choices influence her posture, paying attention to physical holding informs concrete feasible changes to implement in her day. This also breaks a cycle of exhaustion that was feeding obsessive and controlling habits. Developing resources expands Emuru's awareness and provides a safe internal place from which she can meet her suffering. Speaking to her experience and allowing what she says to emerge from a felt sense of the here and now nurtures a greater flow between her physical, emotional and psychological experiences. One step at a time, Emuru feels more at home and at ease with herself and finds a more satisfying way through life.

Through Emuru's story we are reminded that awareness is indeed a matter of the mind. But not of the mind as we most usually conceive it – that is to say our cognition. Rather it is a matter of the mind of our entire body, as Bonnie Bainbridge Cohen explains (2008). This includes our emotional and physical mind as well as our cognitive mind. In fact, the mind of every different fibre and cell of our being. Awareness, understood thus, nurtures a process of becoming that is rooted in the body: it is the fabric of life and health.

To finish, this poem, dedicated to Emuru:

I meet the ground
The ground meets me

In an earthy embrace
I rest

I touch the soul of the earth
My soul awakes
Here I am
Who knew!

Deep down
At the root of my being
A new rhythm
Greets me

New paths unfold within
A fresh way through life
Here, there
I find
Me
Again, and again and again

The words
Foreign at first
Carve
My story

I am
The same
Changed for ever
Unveiled
At last

(Céline Butté, 8.10.2019)

Acknowledgements

This chapter is made public with great sensitivity to my client's anonymity and with their consent. I thank Emuru for the trust in our work, and for the permission to make her unveiling public. Witnessing and journeying alongside her as she softened into the source and expression of her soul was a gift of life.

References

Bainbridge Cohen, B. (2008) *Sensing, Feeling and Action: The Experiential Anatomy of Body-Mind Centering.* (2nd ed.) Contact Editions

Barthes, R. (1977) *Fragments of a Romantic Speech.* Seuil

Bloom, K., Galanter, M. and Reeve, S. (eds.) (2014) *Embodied Lives: Reflections on the Influence of Suprapto Suryodarmo and Amerta Movement.* Triarchy Press

Fischer, J. (2019) 'Sensorimotor Psychotherapy in the Treatment of Trauma', *Practice Innovations* 4(3), 156-165 [online at https://bit.ly/banda28]

Paxton, S. (1987) *Fall after Newton.* Videoda film. Contact Collaborations

Tufnell, M. (2010) *Dance, Health and Wellbeing: Pathways to Practice for Dance Leaders Working in Health and Care Settings.* Foundation for Community Work

---

Céline Butté is a UKCP and RDMP registered dance movement psychotherapist and supervisor, a teacher and a dancer. Her psychotherapy practice spans across twenty years and includes work in mental health, learning disabilities and refugee and asylum seeker services as well as in private practice. She has taught internationally, co-edited *Dance Movement Psychotherapy with People with Learning Disabilities: Out of the Shadows, into the Light* (2017) with G. Unkovich and J. Butler, is a contributing author in *Working Across Modalities in the Arts Therapies: Creative Collaborations* (2018), edited by N. Colbert and C. Bent, and in *Creative Supervision Across Modalities* (2014), edited by A. Chesner and L. Zografou. She has been a dancer for forty years and currently practises Contact Improvisation.

celine@heartofmovement.com ~ www.heartofmovement.com

The Vegetal Body

Plant Body Being: Performing Ecography as 'Alive and Quivering' Stillness

Ali East

Abstract

This chapter proposes humans' genetic plant nature as a fertile source of creative imagination. It suggests that a plant-like awareness can both expand our creative improvisational possibilities beyond the current perceptions of dance and enhance our relationship with vegetal life on the planet. We can align with a 'plantbodybeing' through focused imagery and intuitive dance, which allows us to acknowledge the minimal, subtle yet alive and vibrating stillness similar to that of our plant relations.

As we have thoroughly explored the possibilities of our humanness as dance artists, I propose that we seek out our genetic plant nature as a fertile source of creative imagination. Plant-body awareness implies a "vegetal intelligence" (Chamovitz, 2012) activated throughout our entire skin surface. Current plant biology research has found that plants are acutely aware of their world as in the colour of light, the quality of air, sensitivity to touch and gravity. They are also able to remember and adapt – to pass on, at times, to succeeding generations, traits they have developed in response to environmental circumstances. This is known as an epigenetic process. Plants are instinctive, ego-less, "self-organised and self-created" (Hall, 2011: 13).

In this chapter, I invite you to consider an application of ideas derived from a personal history of eco-choreographic practice. Throughout the chapter I include teaching 'prompts' as succinct, easily absorbed images that need not interrupt, but help to stimulate, both the reader and improviser's flow of ideas.

> From primordial stillness

> Fingers begin to search – a sensuous surfacing along the wooden (it was once a forest) floor. Nails grasp, dig, cling; light enters leaf/skin surface, igniting electrodes, colouring, greening thoughts, signalling kinship through plant/body/being.
>
> (East, 2018)

Vibrating Stillness

> **Prompt:** *Is there a beginning still place that you can find – silent yet alive? Allow a plant-like 'coming to appearance' as a group (as in a forest of diverse species) and without judgement.*

A deep exploration and reflection on our own *awareness* of being most often begins in a quiet still place. I am like a tree – simply present and vibrantly alive. This *definition of bodily awareness* forms a metaphorical bridge between knowing and not needing to know, between the active and the passive, between the vegetal and human worlds. Adopting a *plant-like awareness* allows for a fundamental *sensing, surfacing* and *searching* that expands into a *spiralling, opening, binding and absorbing* with self, others and environment.

Drawing directly from a combination of ecological holism, eco-perception psychology, the physics of organisms, evolutionary palaeontology and plant physiology I suggest that a plant-like awareness can both expand our creative improvisational possibilities beyond the current perceptions of dance *and* enhance our relationship with vegetal life on the planet that is so necessary to our survival as oxygen-breathing beings.

By aligning with '*plantbodybeing*' as dance artists, via imagery and physical interaction, we acknowledge the minimal, subtle yet alive and vibrating stillness of plants. For somaticist Sondra Fraleigh "stillness is bonded to the phenomena of movement, silence and timelessness" (2018: 3). In seeking to express their 'plant-ness', performance artists come to understand an ecological identity that is non-binary (non-gender biased), beneath enculturation, accessible, self-determining, ultimately communicative and democratic (free to express their individual being in their own way within an ecosystem).

I call this work ecography – it brings together an understanding of ecology, environmental science, plant physiology, perception psychology, cognitive science and anatomy through danced metaphor, re-interpretation and intuitive somato-sensory response. The resulting performances may take a variety of forms and happen in a range of formal and informal venues. But they most commonly exist as site-specific, eco-somatic improvisations.

Sharing breath as a metaphor of connection

> **Prompt:** *Can you become a conduit for air (Irigaray & Marder, 2016: 132) channelling it through the body and into the environment? Let your breath move you as you move the air around you.*

Breathing is more than a metaphor of interconnection and interdependence between animals and plants. It is the essence of life, an exchange of air between the inside and out – an invisible, automatic yet (for us humans) controllable action. Using the image of plant breathing we can practise allowing "the breath to pass through us better, with more awareness and attention" (Marder in Irigaray & Marder, 2016: 132). Plant breathing suggests subtle, barely visible and yet vibrantly green or alive stillness.

We depend on trees for our oxygen, our shelter, our very livelihood – and yet, through our greed and ignorance we have destroyed all but small areas of the world's forests to make way for mono-cropping, mining or hydro schemes. Plants and trees are considered economic commodities of mass production. Yet plants are highly intelligent sentient beings capable of communication and feeling. We are genetically related to plants. Many of our fundamental responses, such as our diurnal and nocturnal rhythm, originate from our vegetal cellular memory. Philosopher Matthew Hall comments that "This relational approach stands in contrast to Western treatments of the natural world as a radically different, inferior Other" (Hall, 2011:99). He suggests a return to more 'indigenous lifeways' as offering insight as we search for solutions to the ecological crisis.

For example, in our native Aotearoa New Zealand the indigenous Māori speak of their home ground as their tūrangawaewae – the place where one's feet are woven into the soil as tree roots. Tanē-Mahuta, the atua (god) of the forest is a symbol of shelter, support and inter-connection. In Māori mythology it is the tree god who separated his parents – Papatūānuku (earth mother) and Ranginui (sky father), letting in the light (knowledge) and making space for all living species to exist.

Prompt: Can you visualise a place of origin or belonging – your tūrangawaewae – and begin to weave yourself into that place through movement?

Presence and Expressivity

Prompt: Practise merging with the environment around you while allowing a 'coming to appearance' individually or with others.

As performance artists, we are concerned with qualities of *theatrical presence.* I refer here to a phenomenological intuitive *somatic presence* (Fraleigh, 2000), which I define as involving one's total and simultaneous attention to self, others and environment. It is a state of highly conscious *total awareness* – a kind of presence that is, simultaneously internally focused and outwardly oriented involving an attitude of merging, participating with the world as one egoless empathic self to another. Philosopher Michael Marder describes the kind of presence a tree exhibits as one of "utmost unconscious attention to life itself" (Marder, 2013: 109). He describes a plant-like "coming to appearance" or "coappearing together" (Irigaray & Marder, 2016: 168).

A tree is an expression of form, its "non-conscious intentionality" (Marder, 2013: 37) or unthinking consciousness, that which is peculiar to its vegetal nature, is deeply rooted in its living purpose. One might say that trees simply express their aliveness, their being – through "biochemical signalling and in an incessant wild proliferation" (ibid: 37).

More than anything, trees teach us about presence – about stillness and about *being present* in each moment and through extended time. They tap into our psyche and spirit – the very essence of aliveness. No wonder forests, since our earliest human beginnings, have been designated as sacred with special groves set aside for meditation, prayer and ritual sacrifice. As Marder suggests, "To think and dream with and of plants is to delve into a profound source of thinking and imagination" (Marder in Irigaray & Marder, 2016: 208). Through a danced engagement 'vegetal thinking and dreaming' becomes vegetal *moving and being* – a physical sharing through a trans-conscious awareness as *plantbodybeing*. This practice aims to move students towards a realisation of themselves as part of nature's elemental community through performance of "non-conscious intentionality" that is, according to Marder (2013: 152), non-cognitive, non-ideational and non-imagistic.

Vegetal temporality and visible rhythm

According to Marder (2013: 109), "Vegetal temporality persists in us, despite being modified, internalized, and consequentially concealed". Our concept of extended time-space, our daily and seasonal rhythmic temporality comes from our vegetal DNA. We observe how the infinite repetition and variation of leaf, branch and root structure creates visually rhythmic, repeating patterns of relationship – a vegetal ecography.

> **Prompt:** *Find one shape or simple identifying action motif that can be repeated, varied, adapted or shared to form repeating patterns and rhythms.*

Fig. 1: *Visible rhythmicality or form language.* Photos: Ali East

As part of nature's expressive processes, we share the same potential within our dance-making. Repeated dynamic gesture creates visible rhythmic coherence as movements are shared amongst the group. Like plants, we cooperate with and adapt to the actions of others around us, respond to physical touch, offer support as we navigate time and space together. Like leaves on a tree, no two versions of the action are ever the same. Spatial gestural expression is also the language of plants – wordless, egoless and outward reaching.

> *Prompt: Allow an imitating or adaptation of another's movement into your own as a seamless, unquestioning incorporating into your egoless dance.*

> *Prompt: Play with the counterpoint of similar and different phrasing between your and others' action.*

Timing in a danced ecography creates itself as it happens. Layered time signatures form rhythmic counterpoint – appearing and dissolving from moment to moment. The time of the event itself is open-ended since the dance will find its completion by mutual sensing and agreement. We might make a metaphorical link with the intuitive variations that happen with plants as they enact their individual and collective rhythm of growth and decay daily and seasonally, driven by external environmental conditions and by internalised light energy (photosynthesis). Linear time seems to disappear as they/we live only in the intuitive moment of the dance. The immediate passage of time is different for plants. The time taken for the spiralling circular movement of shoots and tendrils is generally not visible to our eyes yet can be captured on time-lapse photography.

> *Prompt: As a bean tendril can you begin to search outward? – sensing, searching, surfacing, spiralling, surrounding, clinging, holding another body.*

> *Prompt: Pathways overlap and intertwine as an interconnected 'root' system creating visual rhythmic patterns of communication in space/time.*

Plants communicate

Metaphors of plant-like behaviour extend into our exploration of the ways we share information as improvising performance artists. It is now known that plants communicate with each other via "thread-like filaments of fungi that connect roots in complex communication networks" (Devlin, 2018). In the dance, ideas also spread between the performers as if they are linked by some invisible 'rhizomal' thread. A brush, a visual or aural cue, a *matching* or contrasting of form – we slip/fit around the spaces that each other creates, not pausing to plan or consider: rather an instinctual tracking through the space-time as the dance creates itself.

Prompt: Allow your body to slip/fit/reach into the spaces occupied by others in an ongoing delicate surfacing over and around another being or shape in the environment.

Making a direct link with human thought processes Marder explains rhizomatic thinking as "the inextricable relation to 'an outside' – to something other, including parts of inorganic nature, other living beings, and the products of human activity" (Marder, 2013: 168). In the same way, improvisers build on the 'affordances'[1] at hand – their own internal perception, an intuitive action, prompt, sound, object, another body.

Prompt: What are the affordances – the possibilities or potentialities – within your environment? Allow them to shape and re-shape your dance.

Prompt: Allow a rhizomatic exchange of energetic action between bodies and vegetal beings in the landscape.

Fig. 2: *The author performs Plant-body matching.* Photos: B. Snook

In a conscious linking of plant imagery with the human body/soma and the way it interacts with the world, dancer and Shin Somatics practitioner Denise Purvis has coined the term 'rhizo-somatics' to mean the way that all physical, environmental and inter-personal processes are interconnected, communicative and nourishing. She explains, "rhizo-somatic thought and action acknowledges connection to the environment (organic and non-organic), understanding that we are self, but not singular, bringing awareness to our rhizomatic nature of becoming through relationship." (pers. comm.)

Plants Listen

Prompt: Invisible rhizomal filaments connect – convey subtle signals. We listen with every part of our body – allowing each 'sound' to inform the next movement response.

---

[1] In James J. Gibson's words "The affordances of the environment are what it offers the animal, what it provides or furnishes, either for good or ill" (Gibson, 1979: 127).

> **Prompt for the musicians:** *Can you pitch your sounds so that they are received differently by the body. Are there ways to make sound using your own different body parts – voice, feet, etc?*

Just as we respond to the sounds around us, listen to our inner sensory signals, to the "soundwaves [that] vibrate all the tissues of the body" (Olsen, 2002), plants also listen, pick up vibrations and learn from the sound of a chewing insect on nearby leaves or the crackle of fire (Babikova *et al.*, 2013). The image of receiving sound everywhere in the body re-sensitises us to the world. It is another form of vegetal thinking inherited from our plant relations.

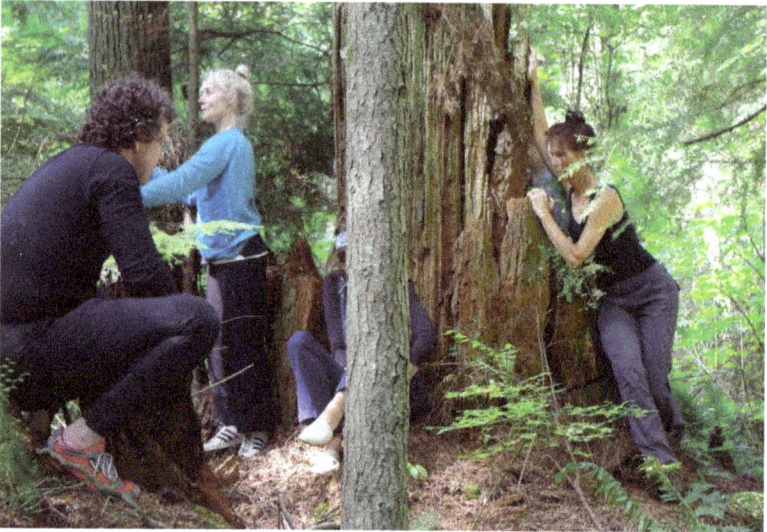

Fig. 3: Plant Listening. Bainbridge Is. Workshop, USA (2019). Photo: Ali East

Performing plantbodybeing:

> **Prompt:** *Trees move/travel through their roots offering possibilities for tethering, nourishment, relationship. Allow an intermingling and overlapping of your bodybeings with other living beings in the space-scape (i.e. the constructed space of the dance).*

Through this kind of prompt, ideas may be inserted into the practice of eco-somatic improvisation in order to search out and reflect on our individual and collective expressive 'nature' (s) – dredge up our plant-like being or (vegetal) consciousness, or awareness, scraping away the surface of habitual behaviour in order to discover what is beneath.

We may begin our practice by considering the ways that the human skeletal structure resembles the spiralling growth in tree trunks: reflecting the way our bones grow in a spiralling configuration and considering the muscles that attach onto the bones in this spiralling formation. Spirals are one of nature's most efficient and repeated patterns.

Or, we might respond to the way that our blood vessels repeat the patterns of leaf veins, and striated rivers (Olsen, 2002; Bainbridge Cohen, 1993).

Or, we might acknowledge how, as plants do, we also *seek out and move towards the light, opening or withdrawing* in reaction to the warmth and light of summer or the chill and darkness of winter. Just as the same phenotypic structures that respond to light and dark exist in both plants and humans (Chamovitz, 2012) plants contain these light-sensitive, energy-absorbing cells throughout their structure. The metaphor of plant-like seeing and light sensitivity across all of our surface area, or skin, changes our orientation as dancers inspiring a multi-directional *cellular sensing and seeing*.

> **Prompt:** *How is my skin's surface intuitively responding to light and shade? How are my feet, fingertips, tailbone etc. seeing/directing/ colouring my dance?*

Or, in another session, we may consider our plant-like 'groundedness' (Bradley, 2009) which implies a bonding with that which is beneath our feet but its importance is as *anchoring* which enables growth, movement upwards. This enables our torsos to fling outwards as a 'counter-tension', to reach in multi-lateral directions, to shudder, rotate or collapse while remaining rooted in place. In Laban's terms, "Stability informs mobility and vice versa" (ibid: 98). In this way we explore our multi-directional, tree-like kinesphere.

> **Prompt:** *From a simple place of standing, let your upper torso reach and twist, extend and spiral in a counter-tension of verticality and horizontality.*

> **Prompt:** *Allow a plant-like reaching out, beyond, towards, around, through the spaces between others. Grounded yet always growing and re-shaping ourselves, amongst others, in this unfinished multi-directional dance.*

All matter responds to gravity, which becomes the key organising principle in moving organisms. Tissues in plant cells contain special properties or cellular protein structures that enable them to grow by spiralling away from gravity (positive gravitropism) while, simultaneously, sending roots downwards (negative gravitropism). We therefore think of a root system anchoring us downwards while sending energy up through torso and limbs from beneath the ground – the image is of *trees in a lake mirror*, as we imagine replicating our size beneath the floor's surface.

In this subtle imperceptibly slow dance they/we move towards, rise or descend through a 'gravitropic' sensing that is acutely present, *aware* and alive.

> **Prompt:** *Allow the presence of others in the space to influence your movement pathway, acknowledging gravitropic pull down into the ground and upwards into the air.*

> **Prompt: Group exercise.** *Beginning from a grounded place allow an extended reaching into space – beyond your own ability to maintain balance. Work in pairs or threes to support someone in this launching outwards as branches on an oak tree. Reaching, falling, recovering, supporting, eventually alternating roles spontaneously.*

We are opening, twisting, inviting our trunks and branches to express – thinking /moving from a grounded place, with multi-directional and multi-dimensional plant-like awareness. All of our skin's surface is sensing, seeing, listening, responding. We acknowledge the dynamic spectrum of stillness and action – internal and external movement that is – one moment minute, cellular, subtle, shimmering and the next moment bold and wild like an exploding seed pod casting its contents into the air.

> **Prompt:** *Within your continuous dance allow a dynamic interactive sharing, fertilising, expanding, dispersing, exploding, re-invigorating and even repelling.*

We respond to environmental stimuli – *the movement we see others make, the image of wind in our branches, the soundscape* played by participating musicians or digitally produced, or to the *patches of light and shade* in the room or forest. We are *surfacing, spiralling, intending towards or away* from objects and each other.

We dance this *form language*[2] (Reigner, 1993; Spirn, 1998) that creates itself spontaneously, intuitively – that becomes its own landscape and part of the psychogeography[3] of the place (Debord, 1955/1989; Eggleton, 2010). Patterns of relationship repeat (Sewell, 1999) overlap, form and reform –

---

[2] Anne Spirn (1998) explains a way of reading the shapes in the landscape natural and man-made (such as plough furrows) as 'form language'.

[3] Psychogeography was defined in 1955 by Guy Debord as "the study of the precise laws and specific effects of the geographical environment, consciously organized or not, on the emotions and behaviour of individuals".

each moment/move informing the next. The dance and dancer appear to dissolve, evolve (create and re-create themselves) through time and space as *a performed egoless ecology of being*. Thinking and non-thinking become one, closely entwined with an external 'other' (Marder, 2013: 164). As poet and philosopher Denys Trussell (2008: 102) contends, we are nature performing ourselves – not as an "author of nature but [as] a potentiality within it". As such, we share the same behavioural phenomena as plants and trees – response to the environment, to weather, to the proximity of other bodies.

To conclude

By aligning with a '*plantbodybeing*' through focused imagery and intuitive dance we acknowledge the minimal, the subtle yet alive and vibrating stillness similar to that of our plant relations. In our hyper-busy human world, such moments of stillness create a dynamic hiatus, pause or ar-rest in the frantic choreography of life. Our performance becomes a practice in *being* – the art of being wholly present to self and others. This suspension of linear time is what cognitive biologist Francisco Varela (1999) called "nowness". Through their artistic explorations, practitioners come to perceive all living species within a relational ecology of "reciprocal responsibility" (Hall, 2011: 115), one of the broader aims of this practice.

Our performance becomes a visual tracing and mapping of branch and limb, root system and trunk – as vegetal and human plant body beings intersect, merge into the one shared ecography. To précis eco-perception psychologist, Laura Sewell (1999: 138), when we invest ourselves, our attentiveness, in the other and surrender our egoic self-centredness the natural world of which we are a part is revealed to us. As Tim Morton (2007) suggests, art can allow stillness yet cause vibration and we know well as dancers that stillness is never static – it is 'alive and quivering'.

Running to the centre
Of this freshly ploughed field
I plant myself firmly in the soil
And begin to grow again
Delicate green.

(Ali East, 1976)

The Vegetal Body

References

Bainbridge Cohen, B. (1993) *Sensing, Feeling and Action: The Experiential Anatomy of Body-Mind Centering.* Contact Editions

Babikova, Z. *et al.* (2013) 'Underground signals carried through common mycelial networks warn neighbouring plants of aphid attack', *Ecology Letters,* 09 May 2013

Bradley, K. (2009) *Rudolf Laban.* Routledge

Chamovitz, D. (2012) *What a Plant Knows: A Field Guide to The Senses.* Scribe

Debord, G. (1955/1989) 'Introduction to a Critique of Urban Geography' in *Situationist International Anthology* (Ken Knabb, trans.) Bureau of Public Secrets

Devlin, H. (2018) 'Plants talk to each other through their roots', *The Guardian.* Online at: https://bit.ly/banda21

Eggleton, D. (2010) Waitaha, 'A Canterbury poem sequence' in J. Stevenson, J. Ruru & M. Abbott (eds.) *Beyond the Scene: Landscape and Identity in Aotearoa New Zealand.* Otago University Press

Fraleigh, S. (2000) 'Consciousness Matters', *Dance Research Journal* 32(1), 54-62

_____ (2018) *Back to The Dance Itself: Phenomenologies of the Body in Performance.* University of Illinois Press

Gibson, J. J. (1979) *The Ecological Approach to Visual Perception.* Houghton Mifflin

Hall, M. (2011) *Plants as Persons: A Philosophical Botany.* Univ. of New York Press

Irigaray, L. and Marder, M. (2016) *Through Vegetal Being: Two Philosophical Perspectives.* Columbia University Press

Marder, M. (2013) *Plant Thinking: A Philosophy of Vegetal Life.* Columbia University Press

Morton T. (2007) *Ecology Without Nature: Rethinking Environmental Aesthetics.* Harvard University Press

Olsen, A. (2002) *Body and Earth: An Experiential Guide.* University Press of New England

Reigner, M. (1993) 'Toward a Holistic Understanding of Place' in D. Seamon (ed.) *Dwelling, Seeing and Designing.* State University of New York Press

Sewell, L. (1999) *Sight and Sensibility: The Psychology of Perception.* Jeremy P. Tarcher/Putnam Press

Spirn, A. (1998) *The Language of Landscape.* Yale University Press

Trussell, D. (2008) *The Expressive Forest: Essays on the Arts and Ecology in Oceania.* Brick Row Publishing

Varela, F. J. (1999) 'Present-Time Consciousness', *Journal of Consciousness Studies,* 6 (2-3), 111-140

Further Reading

East, A. (2015) 'Performing Body as Nature' in *Moving Consciously: Somatic Transformations in Dance, Yoga and Touch*, S. Fraleigh (ed.) University of Illinois Press

_____ (2019) 'Somatic Sensing and Creaturely Knowing in the University Improvisation Class' in *The Oxford Handbook of Improvisation in Dance*, V. L. Midgelow (ed.) Oxford University Press

Gagliano, M. (2015) 'In a Green Frame of Mind: Perspectives on the Behavioural Ecology and Cognitive Nature of Plants' *AoB Plants*, Vol. 7

_____ (2017) 'The Mind of Plants: Thinking the Unthinkable', *Communicative & Integrative Biology*, 10(2)

Maturana, H. and Varela, F. (1998) *The Tree of Knowledge: The Biological Roots of Human Understanding*. Shambala Press

Paturi, F. (1976) *Nature, Mother of Invention: The Engineering of Plant Life*. Harper & Row

Siu, R.G.H. (1976) *The Tao of Science*. MIT Press

Ali East (MPHED) is a New Zealand dance artist and educator, recently retired from The University of Otago and currently adjunct Professor at Tezpur University, Assam. In 1980, she founded Origins Dance Theatre creating more than 25 eco-political works. From 1989-1996 she founded New Zealand's first choreographic BA programme at Unitec, Auckland. She performs regularly with her improvisational collective 'Shared Agendas'.

Ali has published a number of articles and book chapters. Her book *Teaching Dance as if the World Matters: Eco-choreography – A Design for Teaching Dance-making in the 21st Century* was published in 2011. Her current research investigates eco-political and eco-somatic dance processes and intuitive ethnography.

alison.eastnz@gmail.com

Also available from Triarchy Press

A Sardine Street Box of Tricks
Phil Smith & Simon Persighetti ~ 2012, 84pp.
A guide for anyone making, or learning to make, walk-performances.
"a terrific resource...a handbook for making a one street 'mis-guided tour'."
John Davies, author of *Walking the M62*

Embodied Lives
eds. **Katya Bloom, Margit Galanter & Sandra Reeve** ~ 2014, 336pp.
30 movement performers, therapists, artists, teachers and colleagues from around the world describe the impact of Prapto's Amerta Movement on their lives and work.
"Suprapto's work with movement is radically inventive and his dancing a revelation." **Anna Halprin**

Attending to Movement:
Somatic Perspectives on Living in this World
eds. **Sarah Whatley, Natalie Garrett Brown, Kirsty Alexander** ~ 2015, 300pp.
Somatic practitioners, dance artists and scholars from a wide range of subject domains cross discipline borders and investigate the approaches that embodied thinking and action can offer to philosophical and socio-cultural inquiry.

Walking Art Practice:
Reflections on Socially Engaged Paths
Ernesto Pujol ~ 2018, 160pp.
A text for performative artists and cultural activists that brings together Pujol's experiences as a monk, performance artist, social choreographer and educator.

Covert: A Handbook
Melanie Kloetzel and Phil Smith ~ 2021, 110pp.
30 'movement meditations' to diminish the lure of the screens, sidestep invasive scrutiny, and nurture the dialogue between our conscious and unconscious selves.

Stone Talks
Alyson Hallett ~ 2019, 110pp.

The book invites us to listen again to the world around us – the world of rocks, trees, sky, stars and sea that we participate in and that participates in us.

"I will be an avid reader of *Stone Talks*, and I'll tell my friends to keep an eye out for it. I love the way you travel." **Donna Haraway**

Small Arcs of Larger Circles
Nora Bateson ~ 2016, 210pp.

Building on Gregory Bateson's *Towards an Ecology of Mind* and her own film on the subject, Nora Bateson here updates our thinking on systems and ecosystems, applying her own insights to education, organisations, complexity, academia, and the way that society organises itself.

"In this book that moves above all by its questions, Bateson embodies that rarity, a truly free thinker also fully engaged with the fates of all." **Jane Hirshfield, Chancellor of The Academy of American Poets**

Walking Bodies
eds. **Helen Billinghurst, Claire Hind & Phil Smith** ~ 2020, 340pp.

Papers, provocations and actions from the 'Walking's New Movements' conference (University of Plymouth, Nov. 2019)

Guidebook for an Armchair Pilgrimage
John Schott, Phil Smith & Tony Whitehead 2019, 144pp.

"It is wonderful - a brilliant idea, beautifully done, with a sweetly companionable tone to the writing." **Jay Griffiths**

"This mind-pilgrimage is both rural and urban, sacred and neglected, involving the kind of places which, if allowed, can coax the consciousness into insight." **Northern Earth**

walk write (repeat)
Sonia Overall ~ 2021, 92pp.

A handbook that invites readers to use walking as a tool for creative thinking and writing. Offering a whole array of sparks, experiments, projects, catapults, prompts, drifts and exercises, Sonia Overall invites us to see walking as a creative writing method.

Ways to Wander
eds. **Claire Hind & Clare Qualmann** ~ 2015, 80pp.
54 intriguing ideas for different ways to take a walk – for enthusiasts, practitioners, students and academics.

Ways to Wander the Gallery
eds. **Claire Hind & Clare Qualmann** ~ 2018, 92pp.
25 ideas for different ways to walk in and beyond an art gallery – for gallery-goers, walkers, performance artists, students and academics. The book asks us to reconsider our walked relationship with art through the concept of the Wander Score. How playful and embodied can our wandering be in spaces that often make our feet ache?

The Wisdom of Not-Knowing: Essays on psychotherapy, Buddhism and life experience
eds. **Bob Chisholm and Jeff Harrison** ~ 2016, 184pp.
These essays, most by practising psychotherapists, some of them Buddhists, take as their starting point the idea that not-knowing is fundamental to conscious reflection and the desire to know must always arise in the first instance from the self-awareness of not-knowing.

www.triarchypress.net

Also in the series: *Ways of Being a Body*

Volume 1: Nine Ways of Seeing a Body

In the first book in this series, Sandra Reeve succinctly tracks Western approaches to the body from Descartes onwards. The nine ways of seeing a body that she describes are:

~ The body as object ~ The body as subject ~ The phenomenological body
~ The somatic body ~ The contextual body ~ The interdependent body
~ The environmental body ~ The cultural body ~ The ecological body

The book has been very widely welcomed as a guide and stimulus for teachers, students and practitioners.

"I love your book and ...I am now using it as a text in one of my courses."
Don Hanlon Johnson, Professor of Somatics, California Institute of Integral Studies

"...for anyone who has ever trawled through philosophies of the body it is a welcome relief to have them laid out so clearly.
...essential reading for anyone interested in dance, in movement, in philosophies of the body; for dancers, researchers, students, somatic movement practitioners and for dance movement therapists. Wonderful."
Polly Hudson: Senior Lecturer, Dance Performing Arts, Coventry University

"This book is a delightful, readable set of beginning points or lenses through which to constantly consider and reconsider embodied practice."
Phillip Zarrilli: Artistic Director, The Llanarth Group

Volume 2: **Body and Performance**

This edited collection brings together a wide range of contemporary approaches to the body that are being used by performers or in the context of performance training.

The intention is for students, dancers, performers, singers, musicians, directors and choreographers to locate their own preferred approach(es) to the body-in-performance amongst the lenses described here. The collection is also designed to facilitate further research in that direction as well as to signpost alternatives that might enrich their current vocabulary.

All 12 approaches represent the praxis and research of their authors. The chapters reveal a wide variety of different interests but they share the common framework of the notion of 'body as flux', of 'no fixed or determined sense of self' and of supporting the performer's being-becoming-being as a skilful creative entity, emphasising the intelligence of the body at work.

*"Taken as a set of perspectives, this collection has much to offer scholars and practitioners alike, particularly those involved in praxical explorations of embodiment in awareness in performance art or performer training...
...the questions and perspectives included - and subsequently raised - are testament to the strength of the volume."*
Dr Ben Macpherson in *New Theatre Quarterly*

"As a practitioner moving through the book, I was drawn to many new insights made available in detailed accounts of performance praxis previously unfamiliar to me. Indeed, a real strength of this book is its ability to balance a diverse range of approaches by establishing a common point of departure from which each artist's story unfolds. ...The transferable knowledge uncovered between disciplines makes it a highly valuable resource for students, practitioners and scholars alike. With its attractive design and fluid writing style, it is a wonderful follow-up to the first book." Jenny Roche in *Journal of Dance and Somatic Practice*

www.triarchypress.net